ND | Springer Series in **Nonli**

CW00732071

ND | Springer Series in **Nonlinear Dynamics**

Series Editors: F. Calogero, B. Fuchssteiner, G. Rowlands, M. Wadati and V. E. Zakharov

M. V. Nezlin   E. N. Snezhkin

Rossby Vortices, Spiral Structures, Solitons

Astrophysics and Plasma Physics in Shallow Water Experiments

English translation by
A. Dobroslavsky and A. Pletnev

With 88 Figures

Springer-Verlag

Berlin Heidelberg New York
London Paris Tokyo
Hong Kong Barcelona
Budapest

Dr. Mikhail V. Nezlin
Dr. Evgenii Snezhkin
Russian Scientific Center "Kurchatov Institute"
Kurchatov Square 1, 123182 Moscow, Russia

Series Editor:

Professor Dr. George Rowlands
Department of Physics
University of Warwick
Gibber Hill
Coventry CV4
U.K.

ISBN 978-3-642-88124-4   ISBN 978-3-642-88122-0 (eBook)
DOI 10.1007/978-3-642-88122-0

Library of Congress Cataloging-in-Publication Data.
Nezlin, M. V. (Mikhail Veniaminovich) Rossby vortices, spiral structures, solitons : astrophysics and plasma physics in shallow water experiments / M. V. Nezlin, E. N. Snezhkin : English translation by A. Dobrolavsky and A. Pletnev. p. cm. – (Springer series in nonlinear dynamics)
  Includes bibliographical references and index.
  ISBN 978-3-642-88124-4
2. Plasma dynamics. 3. Solitons. 4. Astrophysics. I. Snezhkin, E. N. (Evengiĭ Nikolaevich) II. Title. III. Series. QC159.N48 1993   521 – dc20   93-28898 CIP

© Springer-Verlag Berlin Heidelberg 1993
Softcover reprint of the hardcover 1st edition 1993

Camera ready by author

57/3140-5 4 3 2 1 0 – Printed on acid-free paper

Preface

This book can be looked upon in more ways than one. On the one hand, it describes strikingly interesting and lucid hydrodynamic experiments done in the style of the "good old days" when the physicist needed little more than a piece of string and some sealing wax. On the other hand, it demonstrates how a profound physical analogy can help to get a synoptic view on a broad range of nonlinear phenomena involving self-organization of vortical structures in planetary atmospheres and oceans, in galaxies and in plasmas. In particular, this approach has elucidated the nature and the mechanism of such grand phenomena as the Great Red Spot vortex on Jupiter and the spiral arms of galaxies. A number of our predictions concerning the dynamics of spiral galaxies are now being confirmed by astronomical observations stimulated by our experiments.

This book is based on the material most of which was accumulated during 1981–88 in close cooperation with our colleagues, experimenters from the Plasma Physics Department of the Kurchatov Atomic Energy Institute (S.V. Antipov, A.S. Trubnikov, A.Yu. Rylov, A.V. Khutoretsky) and astrophysics theoreticians from the Astronomical Council of the USSR Academy of Sciences (A.M. Fridman) and from the Volgograd State University (A.G. Morozov). To all of them we wish to express our gratitude. Whenever we speak of "our experiments", the participation of the entire team is implied.

The creation of the English version of the book would have been impossible without the collaboration of Mr. A.S. Dobroslavsky and Mr. A.S. Pletnev.

We hope that the qualitative idiom of this book, tried out at various conferences, symposia and workshops, will be appreciated by a wider audience including research workers, students, and postgraduates interested in nonlinear dynamics and plasma physics, geophysicists concerned with the dynamics of atmosphere and ocean, astrophysicists and astronomers. The book is also expected to be useful to university teachers of physics and related subjects.

Moscow, August 1992

M. Nezlin
E. Snezhkin

Contents

1. Introduction

In today's investigations of natural phenomena, two approaches, apart from analytical theory, play a decisive role: observation and numerical simulation. The latter is based on mathematical and physical models, balanced between simplicity and physical adequacy in approximating reality.

Each of the two approaches has its own clear advantages and well-known shortcomings. For instance, a deficiency of the observational approach is that the observer lacks the means of modifying the natural phenomenon under study so as to facilitate the investigation of its mechanism. Numerical simulations are handicapped, on the one hand, by the limitations of available computing power, often necessitating drastic simplification of the problem, and, on the other hand, by the arising computational instabilities which can be eliminated only by introducing strong numerical viscosity which considerably affects the results, sometimes to the point of utter inadequacy. In this book, we present yet another approach to the simulation of natural phenomena, which may be called the method of analog physical simulation. This method is based on the use of laboratory models, their properties described by exactly the same equations as those for the real object of investigation. This approach has several advantages: first, its experimental implementation is simple and clear; second, the results pertain to a real physical object rather than to its idealized representation in the computer; third, it is free from the shortcomings of numerical simulation mentioned above. Experimental laboratory simulation has already yielded results of fundamental importance which could not have been obtained via computer simulation. On the other hand, the setup of a laboratory experiment becomes much more general and profound when based on theoretical analysis supplemented by numerical simulations. When these two approaches can be pursued independently, they will yield (within a comparable range of conditions) similar results, qualitatively and even quantitatively. Thus it seems worthwhile to combine the two methods so that they stimulate and augment each other. While this book is mainly concerned with the experimental laboratory simulation, we try, whenever possible, to compare the results of these two independent approaches.

The model medium in the experiments described in this book is free-surface shallow water rotating in the Earth's gravitational field. Its thickness, measured along the local normal to its surface, is small compared to the horizontal dimensions of the phenomena under investigation. Two deep physical analogies are associated with this medium. One of them is the analogy between the dynamics of shallow water and of two-dimensional gas. This makes it possible to simulate with shallow water such natural phenomena as the spiral structures of galactic gaseous disks, the largest and longest-lived vortices in planetary atmospheres and oceans (the Rossby vortices). The book shows in detail how this simulation has led to understanding the nature and probable mechanism of these geophysical and astrophysical phenomena and to the discovery of hitherto unknown features. Astronomers were given clues for discovering novel phenomena, and theoreticians received the basis for building a new theory.

When applied to the physics of galaxies, the simulation results provided an experimental foundation for the new hydrodynamic concept of the origin of their spiral structures, which, we believe, might eventually replace the gravitational concept which is now generally recognized. These model experiments have been especially useful because they have stimulated both a new cycle of astronomical observations and a new approach to their interpretation. In observational astronomy and astrophysics, these activities have led to the discovery of hitherto unknown fundamental phenomena which offer a new look at the problem of galactic spiral structures. For example, the most recent observations, carried out in direct connection with and stimulated by the results of laboratory simulation, indicate that – contrary to previous opinion – the majority of spiral galaxies (rather than only a small proportion, as believed earlier) display exactly those peculiarities in their rotation (a velocity jump between the core and the periphery) which constitute the essential part of the experimentally confirmed hydrodynamic concept of spiral structure genesis. Other observational data are in a good agreement with the prediction, made on the basis of laboratory simulation, that intensive vortices are to be found *between* the arms of spiral galaxies.

With regard to the giant planets of the Solar System, laboratory results provided an experimental foundation for the new "solitonic" concept of the nature of large long-lived vortices (the Rossby vortices) such as the famous Jovian Great Red Spot (JGRS), and made it possible to advance the theory in some of its major points. According to this concept (in its new, perfected interpretation), these natural structures can be viewed as nondispersing solitary formations, or vortical Rossby solitons. Simulations have given a clue to understanding the physical cause of the uniqueness of the JGRS over planet perimeter, an age-old mystery. They have also provided a physical explanation of the cyclone-anticyclone asymmetry observed for large Rossby vortices on Jupiter and Saturn: as a rule, the long-lived vortices are anticyclones. A similar cause is apparently behind the observed cyclone-anticyclone asymmetry of frequently encountered large-scale oceanic intrathermoclinic vortices ("lenses"), which for the most part are also anticyclonic. The underlying cause, discovered through laboratory simulation, is that the large Rossby anticyclones are solitons and therefore do not disperse, whereas the cyclones, not being solitons, undergo a rather rapid decay. (We define a vortex to be a cyclone/anticyclone if the vorticity in its central part is parallel/antiparallel to the component of the angular velocity vector of the overall system rotation along the local normal to the free fluid surface.)

As regards the theory of Rossby solitons, the results of laboratory simulation indicate that a Rossby soliton can only exist as a "genuine" vortex – that is, a structure which contains closed streamlines and carries the captured fluid along with it. This fact has radically changed the concept of Rossby solitons, enabled one to explain a number of their novel properties (for instance, their merging upon collision), and was decisive for the creation of the new theory which is based on the vortical property of Rossby solitons as discovered through laboratory simulation.

The other analogy is between drift motions in rapidly rotating shallow water and in magnetized plasma.[1] In particular, Rossby vortices and solitons are physically similar to drift vortices and solitons in magnetized plasma. This analogy allows one to predict, through the observations of Rossby vortices and solitons in rotating shallow water, the properties of drift vortices in plasma which previously could not be detected in theory or in direct plasma experiment. Another advantage of these simulations is that the phenomenon of interest can be studied *in crudo*, while in real plasma its investigation is hampered by the presence of other nonlinear effects. The simulation has offered new insight into the drift vortices and solitons in magnetized plasma, leading to a radical revision of previously existing views. It also helped to predict new phenomena, such as, for example, the merging of monopolar vortical drift solitons and the decay of dipolar drift vortices. The dipolar drift vortex, which had been theoretically described as a nondispersing soliton, was shown to actually decompose, under certain conditions, into a genuine monopolar soliton and a rapidly decaying vortex of opposite polarity.

Applied to geophysical hydrodynamics and meteorology, the shallow water experiments described in this book made it possible for the first time to produce and study in the laboratory Rossby solitons (solitary waves), hitherto known only in theory. The study has shown that the actual properties of Rossby solitons are radically different from the theoretical notions which existed at that time. It was discovered that such solitons must necessarily be "genuine" vortices and, besides, they merge upon collisions (rather than pass through each other, as was formerly assumed), have a different structure, *etc*. Finally, these experiments substantiated the idea that the natural structures under consideration (large Rossby vortices in the atmospheres of giant planets and spiral waves in galaxies) have a common mechanism of generation: the hydrodynamic instability of gas flows with a velocity shear.

This preamble should convince the reader that the hydrodynamic method of physical laboratory simulation, as described in this book, is a promising research technique for studying both astrophysical phenomena and collective effects in plasma physics. Of course, this method would be even more efficient if combined with computer simulation.

---

[1] According to A.M. Obukhov, this analogy was noted in the '60s by M.A. Leontovich.

2. The Natural Phenomena Simulated in Rotating Shallow Water Experiments

In this book we deal with the simulation of large-scale and long-lived structures generated in planetary atmospheres, oceans, and galaxies. In order to outline the scope of the relevant phenomena, we must first define the physical scales of length, proper rotation velocity, and lifetime. Having done this, we shall concentrate our attention on the phenomena of planetary scale (those occurring in planetary atmospheres and oceans). The conditions existing in galaxies will be considered at a later point.

2.1 Length, Rotation Velocity, and Lifetime of the Structures

Observations reveal that planetary atmospheres and oceans contain pronounced large-scale, long-lived structures: zonal flows along the parallels and large vortices. The laws which determine the parameters of these structures, their dynamics, the conditions under which they can appear and exist, are most clearly manifested on giant planets, Jupiter and Saturn in particular, where the crucial scale parameter, the ratio of planet radius to the characteristic horizontal dimension of the structure in question, is large enough (about ten or higher). For terrestrial atmospheric structures this parameter is a few units, which is presumably the reason why such structures are not so well pronounced on Earth. Thus, in order to understand the properties of large atmospheric structures, it would be wise first to study the situation on the giant planets and then come down to Earth. This is, generally speaking, just what we are going to do. First, however, we shall work out the physical criteria which allow one to classify vortices, both in atmospheres and in oceans, as "large-scale" and "long-lived". These criteria provide the basis for distinguishing from among the various observed natural structures the so-called planetary waves and vortices, named after the prominent meteorologist Carl Gustaf Arvid Rossby (1898–1957), who was the first to point out the fundamental role they play in global atmospheric circulation [2.1–5]. Of equal importance is the role of Rossby waves and vortices in ocean dynamics. They are the synoptic (mesoscale) oceanic vortices which carry a considerable portion of the entire ocean's kinetic energy. It is the reinforcement of oceanic flows by Rossby waves that gives rise to global flows such as the Gulf Stream, the Kuroshio current, and others [2.6–8].

There is a clear-cut physical reason that singles out Rossby vortices into a separate class of structures and makes them so important for terrestrial and planetary physics.

First of all, the crucial point is that all the structures of our current concern (for instance, the JGRS vortex whose effective thickness is a few dozen kilometers while its horizontal dimensions are three orders of magnitude greater) can be treated within the framework of the "shallow water" theory. "Shallow water" is the name for a liquid with a free surface, located in a gravitational field, whose thickness (vertical dimension) is small compared to the characteristic horizontal scales pertaining both to motions in the liquid and to bottom inhomogeneities. It is a physically two-dimensional hydrodynamic medium which is in hydrostatic equilibrium with respect to vertical motions – or, more precisely, in which the vertical velocities are negligibly small compared to the velocities of horizontal motions. The latter motions are usually caused in such media by the force related to the gradient of hydrostatic pressure $g\nabla(\delta H)$, arising when liquid layer thickness changes as a result of a deviation of the free surface by a certain value δH from its unperturbed position corresponding to the state of rest (g is the acceleration of gravity; δH is measured along the z axis which is directed upwards along the normal to the unperturbed free surface of the liquid).

Since the planetary structures under consideration exist in a rotating frame, they are influenced by the centrifugal force and the Coriolis force. The first of these two inertial forces, as is generally known, can be expressed as a gradient of some potential and therefore can be easily combined with the force resulting from the hydrostatic pressure gradient [2.9, 10]. As a result, the gravity acceleration g will be replaced by the effective acceleration g^* corresponding to the vector sum of the gravitational and centrifugal forces; motions directed along the surface perpendicular to the effective acceleration vector will be called horizontal in this case. Generally speaking, while the difference between g^* and g is substantial in the experimental configurations described below, on real planets $g^* \approx g$. The other inertial force, the Coriolis force, cannot be expressed as a gradient of a potential. Therefore, it has to be accounted for by a separate term in the differential equation of motion we are now going to consider [2.9, 10].

So, let the free surface of shallow water be displaced by δH from its state of rest due to a perturbation, so that the total (perturbed) layer thickness, measured along the z axis, becomes $H = H_0 + \delta H$ where H_0 is the unperturbed value (for instance, the effective thickness of an atmosphere is $H_0 = T/Mg$ where T is the gas temperature and M is the mass of its molecules). Then the horizontal motion of nonviscous heavy liquid in the shallow water approximation is governed by the well-known Euler equation which in our case has the form [2.9, 10]

$$\partial v/\partial t + (v\nabla)v + (f \times v) = -g\nabla(\delta H) \tag{2.1}$$

(see Supplement S2.1). Here v is the velocity, t is the time, $f = 2\Omega_z$ is the vector Coriolis parameter, Ω_z is the local vertical component of the angular velocity vector Ω of overall system rotation, so that

$$f = 2\Omega \cos \alpha \qquad (2.2)$$

where α is the polar angle between the vector Ω and the z axis (on a planet, this will be the angle complementary to latitude).

Throughout this book we shall deal with such a motion regime of natural vortices where the total derivative dv/dt, corresponding to the first two terms of (2.1), is small compared to the Coriolis acceleration (the third term). This regime is called the Rossby regime and the vortices satisfying this requirement are called Rossby vortices. Mathematically, this condition is expressed by the following inequalities:

$$l \gg v/f , \quad \text{or} \quad \text{Ro} \equiv (v/l)f^{-1} \ll 1 \qquad (2.3)$$

where l is the characteristic horizontal dimension of the vortex and Ro is the Rossby-Kiebel number.

In the Rossby regime, the Euler equation (2.1) reduces to the equation of the so-called geostrophic equilibrium:

$$(f \times v) \approx -g\nabla(\delta H) . \qquad (2.4)$$

The smallness of the Rossby-Kiebel number is the prerequisite for system rotation to have a strong influence on the properties of the relevant structures. Indeed, for the rotational effects to be significant, the time it takes a fluid volume to pass the characteristic length l at velocity v must substantially exceed the period of system rotation, which is actually the meaning of (2.3).

So we shall be interested in the vortices that comply with the Rossby regime condition (2.3) where the Coriolis acceleration is much greater than, say, the relative acceleration due to proper rotation of the vortices (the second term in (2.1)). According to (2.3), such vortices must be sufficiently large. Let us estimate their spatial scale l, considering atmospheric vortices first. Note that in planetary atmospheres all natural motions are subsonic – that is, their velocity does not exceed the isothermal speed of sound $c_s = c_0 \equiv (gH_0)^{1/2}$. Therefore, to satisfy the inequality (2.3), it is sufficient (but of course not necessary) that $l \gtrsim c_s/f$, or

$$l \gtrsim r_R \qquad (2.5)$$

where

$$r_R = \frac{c_0}{f} = \frac{(gH_0)^{1/2}}{f} \qquad (2.6)$$

is the Rossby-Obukhov radius, a natural scale parameter for large-scale dynamics of atmospheres.

At moderate latitudes in planetary atmospheres, the magnitude of r_R is about 3000 km for Earth and about 6000 km for Jupiter and Saturn.[1] Long-lived planetary vortices of such dimensions comply with the Rossby regime condition (2.3) and are the objects of our investigation. According to (2.3), the characteristic periods of their proper rotation $2\pi l/v$ are much longer than the diurnal rotation period of the planet. Thus, we leave out of our consideration the more rapidly rotating vortices such as tornadoes and typhoons. The interested reader is referred to the recently published and richly illustrated books [2.11–16] and to the excellent album on fluid motion [2.17].

Apart from the scale r_R, there is another intrinsic scale r_i, the inner (baroclinic) Rossby radius, in the dynamics of atmosphere and ocean. Unlike the barotropic radius r_R which corresponds to a vertically uniform medium, the scale r_i pertains to a vertically inhomogeneous (stratified) medium [2.9, 10, 18]. The baroclinic Rossby radius is given by

$$r_i = \frac{N H_0}{\pi n f} \tag{2.7}$$

where

$$N = [-(g/\varrho)\partial\varrho/\partial z + g^2/c_s^2]^{1/2} \tag{2.8}$$

is known as the Brunt-Vaissala frequency of vertical oscillations in a medium stable against convection (that is, a medium whose density ϱ decreases with altitude z); n is the vertical mode number (that is, $k_z = \pi n/H_0$ is the vertical wave number).

Now let us explain the physical meaning of the baroclinic Rossby radius r_i. For this purpose, we neglect medium compressibility, that is, the second term in brackets in (2.8). Then for $n = 1$ we obtain from (2.7)

$$r_i = \left(g\frac{\Delta\varrho}{\varrho}H\right)^{1/2} \Big/ \pi f = (g'H)^{1/2}/\pi f \tag{2.9}$$

where $\Delta\varrho = |\partial\varrho/\partial z|H$ is the change in medium density over the distance H and $g' = g\Delta\varrho/\varrho$ is the so-called "reduced acceleration of gravity". Apart from the factor of π, the expression (2.9) for r_i differs from the expression (2.6) for r_R only in that g is replaced by g'. This replacement accounts for the action of the Archimedes force and is always necessary when the motion of a layer of thickness H and density $(\varrho - \Delta\varrho)$ over a thicker layer of density ϱ is considered (see, for example, [2.9]). As to the factor of π, it appears in the equation due to the fact that, as it turns out, all the equations of horizontal motion for a stably stratified medium differ from those for a two-dimensional medium in only one detail: the baroclinic equations contain the inner Rossby radius (2.7) instead of the Rossby-Obukhov radius (2.6) used in the barotropic equations. Thus for

---

[1] The calculation of the Rossby-Obukhov radius for the atmospheres of Jupiter and Saturn is based on the effective thickness of the atmosphere H_0 in its upper cloud layer; thus for Jupiter $H_0 = 25$ km.

our future discussion we note that the scale r_R is to be replaced by r_i in the interpretation of astronomical observations.

The baroclinic radius r_i is always smaller than the barotropic radius r_R. It is several times smaller in the atmosphere and several dozen times smaller in the ocean. For instance, in an isothermal atmosphere, assuming a Boltzmann altitude profile of density, we get from (2.7) and (2.8) for $n = 1$ (we always have in mind the first baroclinic mode):

$$r_i/r_R = \pi^{-1}[(\gamma - 1)/\gamma]^{1/2} \approx 1/6 \qquad (2.10)$$

where $\gamma = 1.4$ is the isentropic exponent. Some typical values are:

$r_i \approx 1000$ km in the atmospheres of Jupiter and Saturn;

$r_i \approx 500$ km in the Earth's atmosphere;

$r_i \approx 50$ km, $r_R \approx 2000$ km in the Earth's oceans (for $H_0 \approx 4$ km).

So in the present book we shall call large-scale such vortices whose dimensions are greater than either r_R or r_i, depending on the actual problem. [Sometimes, vortices whose size is between r_i and R (the planet radius) are referred to as "mesoscale" vortices, and those whose size is less than or close to r_i, as "submesoscale" vortices.]

Let us recall some more facts we shall need in the subsequent discussion. According to the equation of geostrophic (cyclostrophic) equilibrium (2.4),

$$f v_0 \approx g(\delta H)_0/a \qquad (2.11)$$

where v_0 is the amplitude (the maximum absolute value) of linear velocity in the vortex, $(\delta H)_0$ is the free surface displacement amplitude, and a is the characteristic vortex radius. For a cyclone, by definition, the vorticity (that is, velocity curl) in the central part of the vortex is parallel to the local normal component of the angular velocity of planet rotation. Accordingly, the Coriolis force acting on the circular current is directed outwards from the vortex center and $\delta H < 0$ (a depression). For an anticyclone, the vorticity has the opposite sign, the Coriolis force is directed towards the vortex center, and $\delta H > 0$ (an elevation).

Obviously, in the field of the meridionally nonuniform Coriolis force in accordance with (2.2) the vortex must drift along the parallel: this is one of the manifestations of the gyroscopic effect in a rotating system (on a planet). In the northern hemisphere, by virtue of (2.2), the Coriolis force is stronger in the northern part of the vortex (closer to planet pole), so particle path curvatures increase in the northern part of an anticyclone and decrease in the northern part of a cyclone. As a result, in a medium free from other flows and with no depth gradients, a vortex drifts westwards (counter to planet rotation) irrespective of its polarity, whether cyclonic or anticyclonic. Vortex drift velocity is given by $V_{dr} \approx \beta a^2$ where the factor β, for constant atmosphere thickness ($H_0 = $ const), is

$$\beta = \partial f/\partial y = -R^{-1}\partial f/\partial\alpha , \qquad (2.12)$$

R being the planet radius and y the meridional coordinate (increasing in the direction of decreasing α, i.e. northwards in the northern hemisphere). In the so-called β-plane approximation, β is assumed to be a constant, which implies a linear dependence of the Coriolis parameter:

$$f(y) = f_0 + \beta y \tag{2.13}$$

where f_0 is the value of the Coriolis parameter at the longitude under consideration where we assume $y=0$. The mentioned drift is known as β-effect [2.9, 10, 18] and is exactly what gives rise to Rossby waves (Sect. 5.3). As long as vortex radius a is small compared to r_R (or r_i, depending on the actual problem), the drift velocity V_{dr} increases with a. The magnitude of V_{dr}, however, cannot exceed an upper limit which is close to βr_R^2 (or, respectively, βr_i^2) since the so-called Rossby velocity

$$V_R = \begin{cases} \beta r_R^2 \\ \beta r_i^2 \end{cases} \tag{2.14}$$

is an upper limit for the speed at which linear disturbances complying with the Rossby regime (2.3) can propagate in a rotating fluid. (This is discussed in detail in Sect. 5.3.)

As to the oceanic vortices, the motions in them are much slower, roughly by two orders of magnitude, than those in the atmospheric vortices, and the sufficient condition (2.5) is therefore very far from the necessary condition (2.3). Oceanic vortices satisfy the Rossby regime condition (2.3) even when their size is two orders of magnitude smaller than the value stipulated in (2.5). In practice, Rossby vortices in oceans are dozens (sometimes hundreds) kilometers across. These dimensions are close to the inner Rossby radius r_i.

At this point we must emphasize another fact, essentially important in the analysis of nonlinear effects but often overlooked. The fact is that the Rossby-Kiebel number is restricted not only from above (by (2.3)), but also from below. Indeed, Ro is a measure of the nonlinearity inherent in a process or a structure, so it must not be too small if the nonlinearity is to be well developed. For instance, the most interesting – that is, highly nonlinear – Rossby vortex regime is observed when the streamlines contain a separatrix which encloses trapped particles of the fluid carried along by the vortex. As we shall see, for example, in Sect. 5.5, the particle trapping regime sets in only when the amplitude of the linear velocity of vortex proper rotation v_0 exceeds its drift velocity (in the estimates of this kind, we always use the absolute value of the drift velocity):

$$v_0 > V_{dr} \ . \tag{2.15}$$

Because $V_{dr} \approx V_R$, this highly nonlinear regime requires that

$$\text{Ro} > V_R/a\Omega \ , \tag{2.16}$$

that is, since $V_R = \beta r_R^2$, $a \approx r_R$, and $\beta \approx \Omega/R$,

$$\text{Ro} > r_\text{R}/R \ . \tag{2.17}$$

Usually, for naturally occurring conditions, r_i should be used in (2.17) instead of r_R. For instance, combined with (2.3), this will give for the ocean:

$$10^{-2} < \text{Ro} < 10^{-1} \ . \tag{2.18}$$

If a Rossby vortex occurs in a moving medium, it may be, in principle, carried along by the flow and, in particular, may move eastwards. Under special circumstances the vortex may stand still. Then, if it is an anticyclone, a prolonged drought may result such as the summer 1972 drought in the USSR (Sect. 7.6).

The key features of planetary atmospheric circulation are the so-called zonal flows. Today they are considered to be the result of the evolution of quasi-two-dimensional turbulence [2.19–25]. This evolution tends to increase the sizes of merging vortices, whose longitudinal dimensions are, in principle, not restricted, so that closed circular flows are formed (for the oceans, this phenomenon is observed only in the polar regions where the continents are not in the way). However, latitudinal vortex dimensions are restricted due to the nonuniformity of the Coriolis force, and the flows therefore acquire a meridionally oscillating structure: flow direction alternates along the meridian with a characteristic period equal to the so-called Rhines length

$$l_\text{Rh} \approx \pi(2u/\beta)^{1/2} \tag{2.19}$$

where u is the characteristic velocity of the flows. In reality, l_Rh is usually a few times larger than r_R and close to the so-called intermediate geostrophic radius $r_\text{IG} = (R/r_\text{R})^{1/3}r_\text{R}$.

The energy transfer from small to large vortices and to the flows can be regarded as an effect of negative viscosity [2.26]. As shown in Sect. 2.2, zonal flows are especially well pronounced on Jupiter and Saturn. They also exist on the Earth, although they are less regular here. This seems to have something to do with the scale factor (the relatively small radius of the Earth) and with the fact that terrestrial atmosphere is less homogeneous [2.9, 18].

Having established the spatial scales of our vortices, we need to define a time scale according to which vortices will be classified as either long- or short-lived. Since we are studying Rossby vortices, it is natural to base the required time scale on the lifetime of a linear Rossby wave packet τ_l. A linear wave packet is known to decay if there is a dispersion – that is, if the group velocity is not constant over the range of packet wave numbers [2.27]. A packet of Rossby waves is also subject to dispersion which distorts packet shape. As a result, its amplitude decreases and so does the velocity of fluid according to (2.11). Consider a wave packet of Gaussian shape

$$h = \delta H/H_0 \propto \exp(-r^2/a^2) \tag{2.20}$$

where r is the distance from packet center. Clearly, the dispersion spreading time τ_l of the packet should depend on its characteristic radius r_R and on the velocity

V_R, or rather on their ratio r_R/V_R. As will be shown in Sect. 5.4, the characteristic time is indeed proportional to r_R/V_R, the factor of proportionality being strongly dependent on the ratio of the effective packet size a (its "radius") to the radius r_R. At $a = r_R$, the time τ_1 passes through a minimum which in the absence of viscosity is given by $(\tau_1)_{min} \approx 8r_R/V_R$. For atmospheric and oceanic vortices, r_R should be replaced by r_i (see above). Then, making use of the bottom row in (2.14),

$$(\tau_1)_{min} \approx 8/\beta r_i \ . \tag{2.21}$$

Table 2.1 lists the values of $(\tau_1)_{min}$ obtained with (2.21) for the atmosphere of Jupiter (at 22°S, the JGRS latitude) and for the terrestrial atmosphere and oceans (at intermediate latitudes).

Table 2.1. Lifetimes of linear Rossby wave packets

	Atmosphere of Jupiter	Atmosphere of Earth	Oceans
r_i, km	≈ 1000	≈ 500	40–50
β, cm^{-1}s^{-1}	$\approx 4.5 \cdot 10^{-14}$	$\approx 2 \cdot 10^{-13}$	$\approx 2 \cdot 10^{-13}$
$(\eta)_{min}$, days[*]	≈ 20	≈ 10	100–120
[*] Here $(\eta)_{min}$ is the time in which the packet amplitude is reduced to half its initial value. (Terrestrial time units are used throughout the book.)			

The actual lifetime of a linear packet of size $a > r_i$ can be several times longer than the minimum value which corresponds to $a = r_i$. In all events, however, it will be much shorter than the characteristic viscous time (Chaps. 5, 9).

Accordingly, we shall call a Rossby vortex of size $a \gtrsim r_i$ long-lived – that is, nonspreading – if its lifetime exceeds significantly the value given by (2.21). This is one terrestrial year for the Jovian atmosphere, three months for the atmosphere of Earth, and several years for the oceans.

Judging by this criterion, there are no long-lived Rossby vortices in the terrestrial atmosphere, since large cyclones and anticyclones survive only for a few weeks. Some oceanic vortices satisfy this requirement, albeit with no great margin. The really long-lived ones are the Rossby vortices in the atmospheres of Jupiter and Saturn. For instance, the largest atmospheric vortex on Jupiter, the JGRS, has been observed for more than 300 years, which is $5 \cdot 10^3$ times longer than $(\tau_1)_{min}$.

This is a good reason to turn our attention now to long-lived vortices of Jupiter and Saturn.

2.2 Large-Scale, Long-Lived Rossby Vortices in the Atmospheres of Giant Planets. The Cyclone-Anticyclone Asymmetry

The main features of the largest long-lived vortices on Jupiter and Saturn and the zonal flows in the atmospheres of these planets have recently been thoroughly investigated both through Earth-based observations and, most comprehensively, by the Voyager space missions [2.28–48]. These vortices display the following general properties summarized in Table 2.2 and illustrated in Figs. 2.1–5.

Table 2.2. Large, long-lived vortices in planetary atmospheres

Planet	Latitude	Vortex name	Vortex polarity	Lifetime, years	Size along meridian/parallel, 10^3 km	Drift velocity, m/s***	Drift direction	Reference
Jupiter	22°S	Great Red Spot	anti-cyclone	>300	12/25	-3	westward	[2.33]
	19°N	Little Red Spot	anti-cyclone	2–5	5/12	-2.5	westward	[2.35]
	34°S	White Ovals	anti-cyclones	≈50	5/7	4	eastward	[2.33]
	14°N	Brown Ovals ("Barges")	cyc-lones*	>30	1.5/7.5	2.5	eastward	[2.36]
Saturn	75°N	Big Berta	anti-cyclone	>1**	5/7			[2.34] [2.48]
	42°N	Brown Spots	anti-cyclones	>1**	3.3/5	4.5	eastward	[2.34] [2.48]
	24°N	UV Spot	cyclone*	>1**	≈3			[2.34] [2.48]
	55°S	Anne's Spot	anti-cyclone	>1**	≈3		eastward	[2.34] [2.48]

   * With these rare exceptions, all large long-lived vortices are anticyclones. The cyclones of Jupiter and Saturn are usually no larger in size than 1000 km and live no longer than a week [2.36,41].
   ** Observations have begun only recently [2.34].
   *** Drift velocities are given in the so-called System III [2.29,31,32].

(1) Cyclone-anticyclone asymmetry is observed on the giant planets. Nearly all of the long-lived vortices are anticyclones, including the famous JGRS (22°S). The exceptions are the Brown Ovals, or Barges, in the Jovian atmosphere (14°N) and the relatively small vortex UV-Spot in the atmosphere of Saturn (24°N).

(2) There is a hierarchy of vortex sizes and lifetimes. The vortices are several thousand kilometers or more in diameter and can be observed in a more or less stable shape for decades and even centuries. The larger a vortex, the longer its

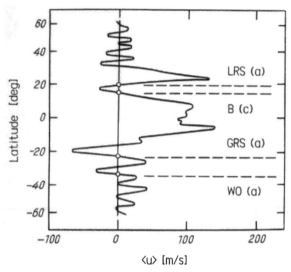

Fig. 2.1. Zonal flows in the upper Jovian atmosphere: average velocity $\langle u \rangle$ of the wind blowing along the parallel (positive wind direction is eastwards) as a function of the planetary latitude [2.35, 39]. The positions of major large-scale, long-lived vortices (*GRS* – Great Red Spot, *LRS* – Little Red Spot, *WO* – White Ovals, *B* – Barges) are indicated as well as their signs (*a* – anticyclone, *c* – cyclone)

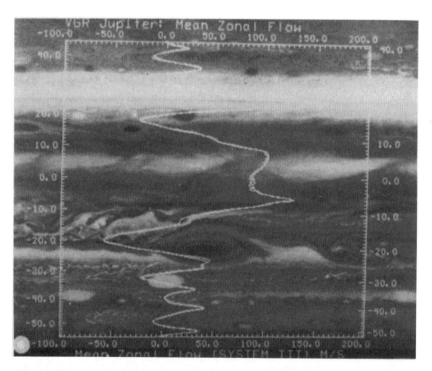

Fig. 2.2. Vortex localization in upper Jovian atmosphere with respect to the meridional profile of zonal flows (*cf.* Fig. 2.1) [2.44]. All vortices except the Barges (14° N) are located in the areas where the zonal flow curl is anticyclonic (that is, they are anticyclones)

lifetime. The longest-lived is the Jovian vortex GRS, known for over 300 years. (First reports on observations of the JGRS and on its westward drift go back to Robert Hooke (1664) and G.Cassini (1665), their observations are described in [2.31, 31'].) South of the JGRS (at 34°S) one finds the Large White Ovals which appeared in 1938 as a result of a strong disturbance in the Jovian atmosphere.

(3) The longitudinal drift of the vortices should be mentioned. All large vortices drift along the parallels: for instance, the anticyclonic JGRS drifts westwards and the cyclonic Brown Ovals drift eastwards.

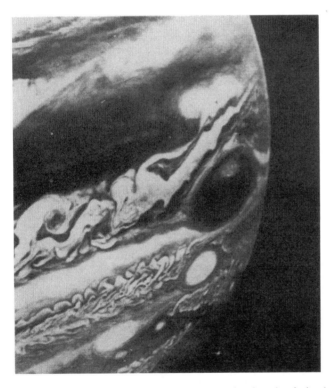

Fig. 2.3. The JGRS vortex in Jovian atmosphere (southern hemisphere), seen as a large dark oval; the White Ovals are to the south and south-west. The zonal structure of winds blowing along the parallels can be discerned, illustration taken from [2.11]

(4) The angular frequency of proper rotation for all large vortices is much lower than the frequency of planet rotation. For instance, Jupiter rotation period is about 10 hours, whereas the characteristic proper rotation period of the JGRS vortex is about a week.

(5) The linear velocities of vortex particles are dozens of meters per second, which is an order of magnitude greater than vortex drift velocities.

(6) The vortices entrain, as they drift along, particles of the medium trapped in them.

(7) Both planets exhibit atmospheric flows (winds), directed along the parallels. Their velocities show meridional oscillation (alternation on Jupiter) with characteristic spatial periods of 15 to 20 thousand kilometers. These are the above-mentioned zonal flows.

(8) All anticyclones are observed in those regions of zonal flows where the vorticity, that is, the flow velocity curl, is anticyclonic.

(9) Cyclones are observed in those regions where zonal flow vorticity is cyclonic. It should be noted that at the latitude of the largest cyclones (Jupiter's Brown Ovals) the flow velocity gradient is very high: the velocity changes by 125 m/s across a meridional distance of as little as 1500 km. The flow velocity gradient is several times higher here than at the latitude of the JGRS.

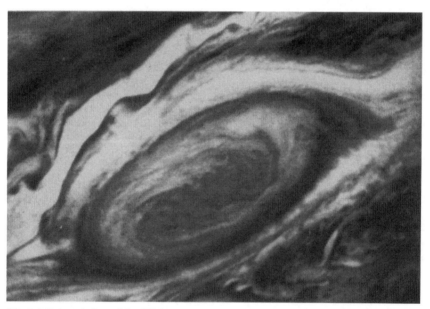

Fig. 2.4. Enlarged view of the JGRS vortex (south is downwards), illustration taken from [2.33]

(10) All large vortices are located at latitudes where the horizontal gradient of flow velocity passes through a local maximum. On Jupiter, for example, the vortices are centered about the line where the flow velocity changes its sign.

(11) The structures in question are conspicuously more or less symmetric with respect to the planet equator. On Jupiter, for instance, the JGRS vortex, drifting south of the equator at 22°S, has a symmetrically located northern counterpart, the Little Red Spot (JLRS) which drifts along 19°N in the same westward direction. The JLRS is morphologically very similar to JGRS (as reflected in its name), the

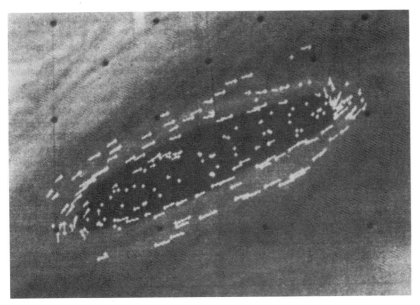

Fig. 2.5. A vortex from the Barges chain on Jupiter and the velocity field within that vortex [2.36]

main differences being that it is 2–2.5 times smaller and has a shorter lifetime of a few years [2.35]. It shows up and disappears with a period of about 5 years. One might suppose that this vortex is marginally stable, and therefore its size is close to the minimum size of a long-lived vortex in Jovian atmosphere.

(12) At the latitudes corresponding to the largest vortices of Jupiter and Saturn, the criterion of hydrodynamic instability for two-dimensional zonal flows is satisfied.

(13) Upon collision, Rossby vortices of like polarities drifting along adjacent parallels tend to merge [2.33].

(14) The vertical dimension of the vortices in question depends on the vertical structure of the planet's atmosphere. The characteristic thickness of the upper cloud layer where, for instance, the JGRS vortex is located is $H_0 = T/Mg$, amounting to a few dozen kilometers on Jupiter and Saturn. This is two or three orders of magnitude smaller than the horizontal dimensions of the vortices. On Jupiter, this thin upper layer rests on a thick and dense layer of gas in which the pressure varies with height by the adiabatic law. This layer is about 1000 km thick. According to the calculations of [2.31], approximately 1000 km below the upper cloud layer, under a pressure of about 5600 atm and at a temperature of about 2200 K, the primary component of Jovian atmosphere, hydrogen, is in the liquid state. In other words, the adiabatic Jovian atmosphere rests upon a liquid "bottom". Still deeper, approximately 20,000 km below the upper cloud layer, under a pressure of about 3 million atm and at a temperature of about 11,000 K, hydrogen is apparently in the metal state. This vertical structure has important

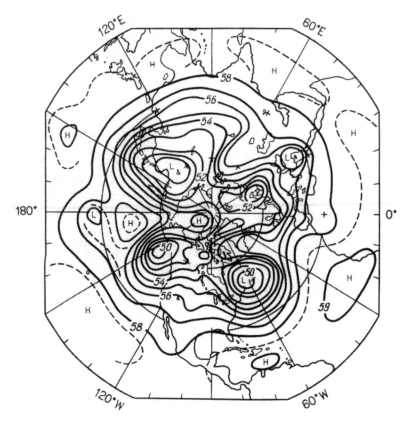

Fig. 2.6. A chain of atmospheric cyclones over the Arctic: equidistant levels of the isobaric surface in northern hemisphere (*L* and *H* indicate cyclones and anticyclones, respectively). From [2.49]

implications for our simulation of natural atmospheric vortices in experiments with rotating shallow water. In calculating the Rossby-Obukhov radius r_R we use $H_0 = 25$ km for the effective thickness of the upper cloud layer in Jovian atmosphere.

It is clear from the comparison of the quoted values of r_R and r_i with the above description of Table 2.2 that the long-lived vortices on Jupiter and Saturn are within the Rossby regime given by (2.3). Their horizontal dimensions exceed the baroclinic Rossby radius r_i, the size of the JGRS exceeding even the Rossby-Obukhov radius r_R. The vortices whose proper rotation is slower than system rotation as a whole (that is, those which rotate in the Rossby regime) and whose dimensions exceed r_R or r_i but are still small compared to the planet radius shall be called – in accordance with the arguments of the previous section – large-scale Rossby vortices.

At this point, it will be worthwhile to draw the reader's attention to the following fact. The properties of giant planet vortices could have been established with the certainty indicated in this section only after the observers had adopted an adequate reference frame – the so-called "System III", mentioned in Table 2.2. This frame is associated with the electromagnetic radiation of the planet (Jupiter, Saturn, etc.) which has a well-defined pattern, making it a very reliable reference. Since this radiation is the result of processes occurring in the planet *core* (like the hydromagnetic dynamo effect), it may be taken as a fixed frame for atmospheric phenomena. The earlier observations, which did not have such a reliable reference system, were indecisive and contradictory, because vortex drifts, for instance, had to be studied, in fact, with a reference to the vortices themselves [2.31].

Examples of large vortices in the terrestrial atmosphere are shown in Figs. 2.6 and 7 where chains of cyclones are seen in the Arctic [2.49] and Antarctic [2.50, 51]. For more details on large terrestrial atmospheric vortices we refer the reader to [2.51–54]. The lifetimes of such vortices, as already mentioned, are short – not longer than a few weeks. Accordingly, there is no use looking for the cyclone-anticyclone asymmetry among the terrestrial atmospheric vortices: this concept makes sense only in the case of long-lived vortices (as defined above). This difference in the conditions on the Earth and on the giant planets (that is, the nonexistence of long-lived terrestrial atmospheric vortices) is due to the fact that Earth surface curvature is too large. Indeed, the ratio r_R/R is less than 0.1 for the giant planets whereas for the Earth it is about 0.5. It is also possible, as indicated above, that this peculiarity of the Earth is due to the strong vertical inhomogeneity of its atmosphere.

2.3 Rossby Vortices in the Oceans

To begin this section, it should be noted that there are global vortices in the oceans, the size of the entire ocean. These are actually zonal flows whose northern and southern streams close in on each other because of the continents. A zonal flow in its pure form can be found only where there are no continents; such is the Antarctic Circumpolar Current (West Wind Drift).

Vortices of this kind are subject to the β-effect which causes Rossby waves and vortices to propagate (to drift) westwards. This leads to the so-called westward intensification of flows – oceanic flows, in this case. Currents can be said to gain additional westward momentum from Rossby waves. Because of the gyroscopic effect (or, equivalently, the Coriolis force), this gives rise to such currents as the Gulf Stream and the Kuroshio [2.6–8]. Instabilities in these currents produce the so-called rings which are a kind of synoptic vortices. These are discussed in the following section.

Here we must point to a nontrivial circumstance: the generation of large vortices by flow instabilities coexists with the inverse process of flow pumping by the ensemble of smaller vortices. This process takes place both in the ocean [2.55–57] and in the atmosphere [2.26]. In the ocean, it is responsible for the

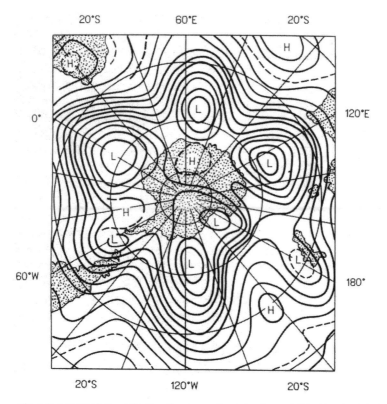

Fig. 2.7. A chain of atmospheric cyclones over the Antarctic: equidistant levels of the isobaric surface in southern hemisphere (L and H indicate cyclones and anticyclones, respectively). From [2.51]

coexistence of currents like the Gulf Stream and rings; in the atmosphere, it is directly related to the dynamic equilibrium between the turbulent cascade which gives rise to zonal flows and the generation of vortices by flow instabilities.

Oceanic vortices, as indicated above, carry a large share of the ocean's kinetic energy. Let us now discuss the principal types of these vortices.

2.3.1 Open Sea Vortices

Vortices of this type are cyclones and anticyclones with horizontal dimensions (radii) of 50–100 km, which are greater than the baroclinic Rossby radius $r_i \approx$ 40–50 km [2.55–57]. Their proper rotation is slow compared to that of the Earth (Ro $\ll$ 1), and they drift along the ocean surface with a westward velocity component close to the Rossby velocity βr_i^2. These properties allow one to classify them as Rossby vortices.

An example of large anticyclones generated by strong winds off the Pacific coast of Central America is shown in Fig. 2.8. Their sizes are several times as large as the local radius r_i. The linear velocities of liquid in the vortices (the

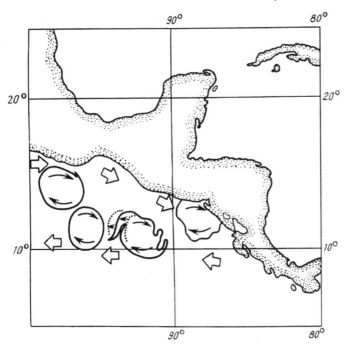

Fig. 2.8. Anticyclones and large-scale currents in the Pacific near Central America in late February, 1976. The anticyclones are outlined along the boundaries separating the warmer surface water outside the vortices from the colder water inside. Westward drift is indicated schematically for one of the vortices. From [2.55]

highest value over the profile) v_0 are several times greater than the drift velocity $V_{dr} \approx V_R$. Thus the Rossby-Kiebel number, in accordance with the criteria (2.3), (2.16) and (2.17), satisfies the condition (2.18).

The parameters of these vortices are vertically nonuniform. For example, the parameter $\varepsilon \equiv v_0/V_{dr}$ varies in the following ranges: for anticyclones, 3–8 near the surface and 2–3 at the depth of 1400 m; for cyclones, 2–7 and 1–3, respectively. The drift velocities of the vortices are also considerably different at different depths. This circumstance, as well as the strong nonlinear interaction between the vortices, restricts their lifetime which is usually no greater than a few months. Consequently, this lifetime does not differ much from that of a linear Rossby wave packet and so the vortices cannot be regarded as long-lived.

2.3.2 Rings of the Gulf Stream and the Kuroshio Current

Observations point to the hydrodynamic instability of the oceanic currents of the Gulf Stream type. The meandering of the currents caused by their instability is so strong that closed loops are formed, detaching themselves from the current [2.55–58]. The vorticity in the detached loops (known as rings) has opposite signs at the opposite sides of the maternal current. For instance, since the Gulf Stream flows from south-west to north-east, the rings south of the current are cyclones and those north of it are anticyclones. The propagation conditions are not the same for rings of opposite signs. The cyclones move thousands of kilometers to the south-west during their lifetime; the westward component of their drift velocity is close to the Rossby velocity. They usually end their existence by merging to the Gulf Stream. The anticyclones, for geographic reasons, join the Gulf Stream much sooner and their lifetime is cut short. The location of the Gulf Stream rings is shown in Fig. 2.9.

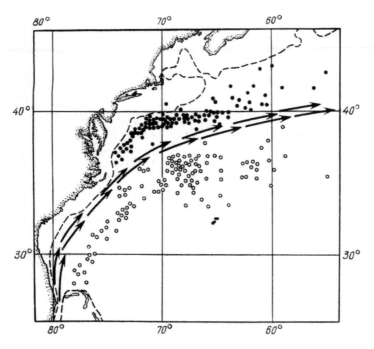

Fig. 2.9. Centers of cyclones (*open circles*) and anticyclones (*filled circles*) of the Gulf Stream in 1967–76. *Arrows* indicate the average location of the Gulf Stream. From [2.55]

Ring dimensions and drift velocities are about the same as those of the open sea vortices; their proper rotation, however, is several times faster (ε =20–30). The rings also satisfy the criterion (2.18). Therefore the rings are Rossby vortices, too. Their lifetimes range from a few months to two years, the upper limit being much above the lifetime of a linear wave packet (Table 2.1).

A comprehensive comparison of these lifetimes was carried out in [2.59] where lifetimes of cyclonic rings were determined by means of numerical simulation, using a mathematical ring model (viscosity neglected) consistent with the observed data. The results are shown in Table 2.3 which gives the calculated lifetimes τ for vortices of different sizes a and different initial amplitudes which are proportional to v_0. The lifetime here, as in Table 2.1, is the time required for ring amplitude to fall to half its initial value.

Table 2.3. Lifetimes of cyclonic vortices of the Gulf Stream, days. From [2.59]

$\epsilon = v_0/V_{dr}$	≈ 0	8	13	27
$a = \sqrt{2}r_i$	65	94	152	536
$a = 2\sqrt{2}r_i$	174	268	532	
For $a \approx 4\sqrt{2}r_i$, the lifetime of a cyclonic vortex is virtually the same as that of a linear Rossby wave packet (2.21); the value of a is found from (2.20)				

We see from Table 2.3 that if the radius of a cyclonic ring is $a = \sqrt{2}r_i$ then with a small amplitude ($v_0 \rightarrow 0$) it lives for about two months. Consequently, its decay is brought about by dispersion, since, according to Table 2.1, its lifetime is the same as the dispersion spreading time τ_1 of a linear Rossby wave packet with $a = \sqrt{2}r_i$. Vortex lifetime increases with its amplitude. This increase, however, is not too significant as long as the velocity of liquid in the vortex remains within an order of magnitude above the Rossby velocity. For instance, if $\varepsilon \approx 10$ then vortex lifetime is just twice that of a linear Rossby wave packet. If, however, the ratio ε is several dozen (while $a = \sqrt{2}r_i$), the ratio τ/τ_1 increases considerably: for example, $\tau/\tau_1 \approx 8$ for $\varepsilon = 27$. [The value of $\varepsilon = 27$ for the case $a = \sqrt{2}r_i$ is the limit: the proper rotation of the cyclone becomes so fast then that the thickness of the upper layer which we consider in the two-layer ocean decreases almost to zero. For a vortex which is twice as large ($a = 2\sqrt{2}r_i$), the limit will be half the above value ($\varepsilon \approx 13$), as it follows directly from the geostrophic equilibrium equation (2.11). Accordingly, the effect will also be weaker: a Gulf Stream ring with $a = 2\sqrt{2}r_i$ and $\varepsilon = 13$ has a lifetime which is only three times greater than that of a linear Rossby wave packet (Table 2.3).] Such rings may well be termed long-lived, unlike the open sea vortices described above whose nonlinearity indicator ε does not exceed a few units and, correspondingly, $\tau \approx \tau_1$. This result fits in well with the observed fact that for relatively small cyclonic rings the nonlinearity indicator ε is several dozen and they survive for much longer than the open sea vortices whose indicator ε is of the order of unity. Nevertheless, even the strongest cyclonic rings suffer dispersion decay, although due to their fast rotation their lifetimes can be prolonged by an order of magnitude. The factors slowing down the dispersion decay of cyclonic vortices will be discussed in Sects. 5.5, 10.3.

Since, for geographic reasons, there are no long-lived anticyclonic rings of the Gulf Stream, only cyclones have been considered in [2.59]. As a matter of fact, the difference between cyclones and anticyclones can be quite critical, as has been pointed out in Sect. 2.2. This will be discussed in detail later, mainly in Sect. 10.3.

Although we have been considering only the Gulf Stream rings here, the same mechanism governs ring generation in other currents: Kuroshio, East Australian, Brazilian, Falkland, and Agulhas. Their rings have practically the same properties as the rings of the Gulf Stream.

2.3.3 Internal Thermoclinic Vortices (Lenses)

Very common oceanic structures are the so-called internal thermoclinic (or intrathermoclinic) vortices, which are found a few hundred meters below the surface, at the horizontal boundary between layers of different salinity, temperature, and density [2.50, 57, 60–68]. They occupy, for example, about 20% of the Arctic Ocean area. These vortices have the shape of lenses; they are smaller in radius than the Gulf Stream rings, though still larger than the local value of r_i, and the westward component of their drift velocity is close to the Rossby velocity. They also comply with condition (2.18). Hence these structures, too, may be classified as Rossby vortices. In contrast to other open sea vortices (not associated with interlayer boundaries), and unlike such vortices as the Gulf Stream rings, the lenses are generated by local perturbations in the ocean rather than by wind and currents. They propagate practically freely, carrying the entrained water over thousands of kilometers. An example showing the propagation of such a lens is given in Fig. 2.10, borrowed from [2.62]. The lenses are spinning fast: the value of ε may be up to several dozen. A highly distinctive feature of these lenses is their cyclone-anticyclone asymmetry: almost all of them are anticyclones. This feature (together with their size in terms of r_i and their drift velocity in terms of V_R) makes them definitely similar to the largest vortices of Jupiter and Saturn. The intrathermoclinic vortices are long-lived: their lifetimes, reaching 10 years and more [2.66, 68], are the longest among the oceanic vortices we have discussed, exceeding the dispersion decay time of a linear Rossby wave packet by more than an order of magnitude (Table 2.1). This corresponds well to the high values of their nonlinearity indicator ε. It must be noted, however, that while the values of their ε are the same as those for the Gulf Stream rings, their lifetimes are several times longer (in this comparison, we take into account not only the actual lifetimes but also the calculated times listed in Table 2.3).[2] This suggests that the difference is due to the opposite polarity of the vortices: the rings are cyclones and the lenses are anticyclones. It follows from both the theory and the simulation experiments that lens lifetimes are apparently limited by viscosity alone [2.64, 65]. The reasons for the longevity of these vortices and for their cyclone-anticyclone asymmetry will be discussed in Sect. 10.3.

To summarize:

1. The lifetime of an oceanic Rossby vortex of either sign depends to a large extent on the degree of nonlinearity ε, defined as the ratio of the amplitude of the linear velocity over vortex profile v_0 to vortex drift velocity $V_{dr} \approx V_R$. When $\varepsilon \approx 1$, vortex lifetime is about the same as that of a linear Rossby wave packet, that is, such a vortex is not long-lived.

---

[2] The Gulf Stream rings are carried off by deep undercurrents in the southern direction where the Coriolis parameter becomes smaller. In accordance with the Rossby-Örtel theorem (see (5.13) below), this tends to intensify the vortices. For this reason the observed lifetimes of the Gulf Stream rings are longer than they would be if they were drifting freely. For anticyclonic lenses, the situation is opposite: their southward deviation weakens them, shortening their lives.

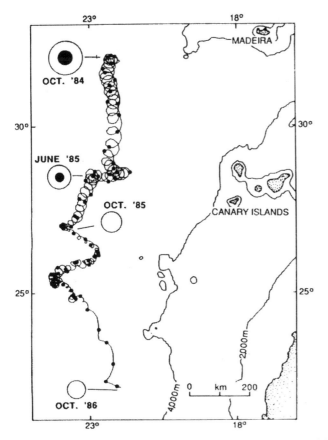

Fig. 2.10. Trajectory of an intrathermoclinic anticyclonic lens in the Atlantic. From [2.62]

2. When the value of ε reaches a few dozen, the lifetimes of cyclonic vortices increase within an order of magnitude but are still substantially restricted by vortex dispersion spreading.

3. In a similar situation (when ε is several dozen), the lifetimes of anticyclonic vortices exhibit a much greater increase; they are now restricted by viscosity rather than by dispersion. This radical difference between the lifetimes of Rossby vortices with opposite polarities is a manifestation of the cyclone-anticyclone asymmetry in the oceans, which is probably similar to the asymmetry observed in the atmospheres of the giant planets.

Before continuing, we would like to make some comments to illustrate our general viewpoint, not going for the time being into a detailed interpretation of these phenomenona.

We analyze large-scale coherent structures in planetary atmospheres and oceans within the framework of a unified concept based on contemporary geo-physical hydrodynamics. The foundation of the concept is the theory of shallow water, whose adequacy to the conditions existing in nature has been substantiated in great detail by Williams in his review [2.69]. This approach allows us to explain the principal properties of large atmospheric vortices: their dimensions, drift, cyclone-anticyclone asymmetry, stability, immunity to dispersion spreading, the mechanism of their generation, their localization at particular latitudes, etc. For instance, it becomes possible to understand the physical reason underlying the uniqueness of the anticyclonic JGRS and the nature of its distinction from the chain of cyclonic Brown Ovals (Barges). Explained is also the coexistence of these (highly coherent) structures with turbulent processes in the atmosphere. These processes result in the merging of vortices with $a \lesssim r_i$ into zonal flows, which turn out to be unstable with respect to the generation of vortices with $a > r_i$.

2.4 Spiral Structures in Galaxies

Most galaxies, including the Milky Way, have well-defined spiral structures like those shown in Fig. 2.11. According to contemporary views, such galaxies are built up of three principal components: a disk, a near-spherical halo, and a corona (Fig. 2.12). We are interested in the disk component – a mostly flat, round formation whose thickness is two orders of magnitude smaller than its radius. The disk comprises a very small fraction of galaxy mass. A small portion of disk mass (about 10% of the Galaxy) is in the form of gas, but this is practically all of the gas available in the galaxy.

It is in the disk that the spiral arms are located, taking up a considerable part of its area. We see the arms as regions of higher luminosity. They are brighter than the rest of the disk not because there are stars within and no stars without: the difference in stellar density within the arms and outside them is as little as 10%. The relatively high brightness of the arms is due to the fact that the process of star formation is more intense here, and newly-born massive stars are much more brilliant than older ones. (There is even a method of evaluating star ages based on their brightness and their emission spectra. By the way, our Sun is a typical old star, aged about 4.5 billion years.) The light of the arms is emitted from excited and ionized hydrogen, as well as from "warm" clouds of molecular hydrogen. Observations indicate that nearly all of the gas is concentrated in the arms; it is the higher gas density in the arms that facilitates the continuing star birth [2.70].

Galactic disks are rotating, and that rotation is highly differential: the angular velocity of matter depends strongly on the distance from the center of the galaxy. An example showing a radial profile of azimuthal linear velocity w_φ typical for spiral galaxies is given in Fig. 2.13. The profile of galaxy rotation depends on the distribution of gravitating masses. The central part of the disk has a relatively compact and dense bulge (Fig. 2.12); the gravitational potential due to the bulge increases from bulge center to its periphery and then falls off rapidly against the

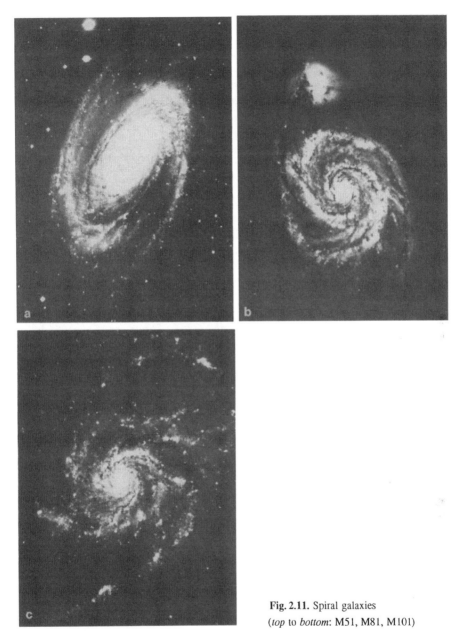

Fig. 2.11. Spiral galaxies
(*top* to *bottom*: M51, M81, M101)

background of the slowly increasing gravitational potential which is due to the rest of the disk, the halo and the corona. The resulting gravitational potential is such that the equilibrium linear velocity as a function of radius (the so-called rotation curve) is represented by an N-shaped curve: the central core is rotating practically as a rigid body and next to it there is a relatively narrow region where the velocity falls off rapidly (Fig. 2.13). To give a notion of the scales of these phenomenona, one should say that ϱ_0 is typically several kiloparsec and w_φ (at its maximum near the center and at the off-center plateau) amounts to hundreds of kilometers per second.

To a first approximation, the rotation of the disk in a spiral galaxy can be thought of as steady, the centrifugal force being counterbalanced by the radial component of gravity. In other words, gravity ensures only the equilibrium of the galaxy.

Doppler observations indicate that the arms of typical spiral galaxies bend backwards, counter to disk rotation ("trailing spirals"). The opposite case of "leading spirals" is never observed in isolated galaxies and occurs very rarely in systems of interacting galaxies, under rather special conditions.

At first glance, the spiral appearance of the galaxies seems intuitively understandable: by virtue of angular momentum conservation, any radial displacements of matter should give rise to spirals. If, however, the spiral arms were due to radial displacements of particles, they would be smeared after a few revolutions because of the highly differential character of galaxy rotation. Indeed, any arm of this kind will be smeared because its central part is rotating much faster than its periphery. According to astronomical observations, a spiral galaxy completes a revolution in about 100 million years, whereas its spiral pattern persists for a hundred times longer. These facts inevitably suggest that galactic spiral patterns are rotating as rigid bodies.

Major difficulties in the interpretation of the fact that spiral structures are rotating as rigid bodies despite the differential rotation of the galactic matter led Jeans [2.71] to suggest that the spiral galaxy phenomenon is a manifestation of some new and yet unknown laws of nature. The mystery of the spiral structure has been tackled by Heisenberg and Weizsäcker, by Chandrasekhar and Fermi [2.72, 73]. A major breakthrough in solving this problem was made by Lindblad in 1941 [2.74]. He suggested that the spiral arms are density waves and their constant shape means that density waves travel about the galactic axis at a constant angular velocity.

Lindblad's wave theory immediately removed the question about "new laws of nature" contemplated by Jeans. Certainly, it did not solve all the problems. Further development of this theory showed that the main problem was associated with the mechanism of wave generation and with the medium in which these waves were supposed to arise and travel.

For a number of reasons, Lindblad's theory was rediscovered a quarter of a century later by Lin and Shu [2.75]. Like Lindblad, they assumed that the density waves are gravitational in their origin and that they arise and propagate in the stellar subsystem of the galactic disk. In the twenty years that followed, no general understanding had been achieved of a universal mechanism for density wave generation and this strongly undermined the theory. Another disadvantage of this theory is the use of an *ad hoc* adjustable parameter which must be specially chosen for each galaxy so as to reconcile theory with observations. Among the

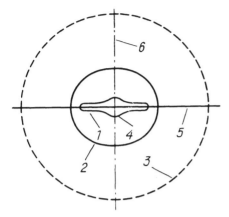

Fig. 2.12. Axial cross-section of a spiral galaxy (schematic): (**1**) disk; (**2**) halo; (**3**) corona; (**4**) bulge; (**5**) central plane; (**6**) rotation axis

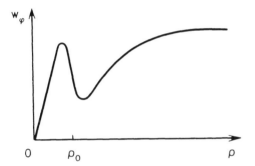

Fig. 2.13. N-shaped rotation curve of a spiral galaxy: linear azimuthal velocity w_φ as a function of the off-center distance ϱ (ϱ_0 is the radius of the velocity jump)

less conspicuous faults is the inability to incorporate galaxies which have an odd number of spiral arms. For details on the gravitational theory the reader is referred to [2.70, 76–80]; an up-to-date criticism can be found in [2.81], see also Supplement S8.1.

The difficulties of the gravitational theory stimulated the birth of a new – hydrodynamic – concept, put forward and developed by Fridman [2.82], see also Morozov [2.83, 84]. Similar to the gravitational theory, the hydrodynamic approach is based on the idea that the spiral arms are density waves in the galactic matter. However, the new concept assumes that the density waves have a hydrodynamic origin and propagate in the gaseous subsystem of the disk rather than in the stellar one. The hydrodynamic concept relies on two observational facts. First, astronomical observations point to a fundamental role played by the gaseous subsystem of the galactic disk in the latter's "spiral" dynamics. For instance, the ratio of gas surface density (the gas being concentrated mainly in the arms) to stellar surface density in the Galaxy is so large that the gas located in the arms almost entirely determines the perturbed gravitational potential [2.85, 86]. Second, in many if not most of the spiral galaxies such as the Milky Way, the Andromeda Nebula, M81, NGC 4594, the rotation curve displays a jump, that is, a region where the azimuthal linear velocity drops abruptly, as mentioned earlier (Fig. 2.13) [2.87–91]. This jump may lead to a hydrodynamic instability of the gaseous component and, as a result, to the emergence of spiral density waves in the gas. The instability is fostered by the massive stellar component of the entire galaxy, which maintains the specific radial dependence of the gravitational potential. In this way, the hydrodynamic theory offers from the very outset a universal mechanism for spiral structure generation. A major advantage of this approach (as opposed to the gravitational theory) is the absence of any adjustable parameters: all characteristics of the spiral structure are unambiguously determined by the rotation velocity profile and the speed of sound in the gas.

The theory confirms that spiral wave generation due to a velocity jump in a differentially rotating gas is indeed possible and that this kind of mechanism is generally more efficient than the gravitational one: the hydrodynamic instability has a higher growth rate, it develops at shorter wavelengths (in particular, shorter than the minimum Jeans length which is the lower limit for the onset of gravitational instabilities), and it is less prone to stabilization by velocity dispersion. A more complete, though also linear, hydrodynamic theory is given in [2.92]. It takes into account not only the radial gradient of rotation velocity but also the radial gradients of gas density and sound speed.

To conclude this section, two important observations should be pointed out. The first is that galactic spiral patterns are almost always azimuthally asymmetrical. The second is that the spiral arms exhibit outward branching in most galaxies. Both these trends are very common in nature, whereas the regular (canonical) spiral structures usually involved in theoretical speculations are exceptionally rare. The whole weight of statistics suggests that these morphological features should be explained not by some random effects but rather by a general cause pertaining to the generation mechanism of the spiral structure. Such a cause was elucidated in the experiments with differentially rotating shallow water and will be discussed in Sect. 8.4.

Additional explanations for some of the material discussed in this chapter are given in Supplements S2.1, 8.1, 10.1.

3. Common Features
of the Simulated Natural Phenomena

According to astronomical observations, natural structures which at first sight are very different, such as the spiral arms of gaseous galactic disks and the large Rossby vortices in the atmospheres of giant planets, exhibit many common features. Their similarity is the theme of this chapter.

3.1 Quasi-Two-Dimensionality

The findings of both Earth-based observations and the Voyager space missions in 1979–82 have revealed a distinct regularity. Despite the considerable vertical nonuniformity of the atmospheres of Jupiter and Saturn, the locations of all the major vortices correlate unambiguously with the *horizontal* structure of the zonal flows, as verified by the following observations (Figs. 2.1, 2). First, the vortices are found primarily at those latitudes of Jupiter where the horizontal velocity passes through zero. Second, the generation of the largest Jovian anticyclones (the Great Red Spot and the Little Red Spot) is nearly symmetrical with respect to the equator, and the vortices are morphologically similar to each other. This suggests that such vortices may be regarded qualitatively as two-dimensional structures. At the same time, calculations of their quantitative parameters should, of course, take into account the vertical nonuniformity of the atmosphere, that is, the difference between the baroclinic radius r_i and the barotropic radius r_R.

Another major point is that spiral arms in gaseous galactic disks can also be viewed as two-dimensional gas dynamics structures, described by the equations of two-dimensional hydrodynamics [3.1, 2]. At first glance, this might seem peculiar, since in a gaseous galactic disk instead of particles we have molecular clouds whose motion proceeds practically without collisions and, therefore, would ostensibly not fit into the hydrodynamic description. Such a description can, however, be justified: the situation here is rather similar to the so-called collisionless hydrodynamics in magnetized plasma. Classical hydrodynamics is appropriate whenever particle motion is isotropic. The motion of neutral particles becomes isotropic due to their collisions with one another. In a strong magnetic field, however, charged particles move in Larmor orbits because of the Lorentz force. Thus their motion perpendicular to the magnetic field is made isotropic by the Lorentz force, and the Larmor radius becomes the equivalent of the free-path length. This justifies the use of collisionless magnetic hydrodynamics for describing the motion of plasma perpendicular to a strong magnetic field, provided that the motion is slow with respect to the Larmor frequency and large-scale with respect to the Larmor radius. This has been shown by Chew, Goldberger and Low [3.3] some thirty years ago (the CGL theorem). In gaseous galactic disks, because the system is rotating as a whole, the motion of neutral particles is influenced by the Coriolis force. Therefore, the particles move in epicycles similar to the Larmor orbits of charged particles in magnetized plasma. The epicycle radii are small compared to disk radius and spiral arm length. As a result, in the case of galactic disks, by analogy with the CGL theorem, the description in terms of two-dimensional hydrodynamics is valid for the motion of gaseous "particles" perpendicular to the rotation axis as long as the motion is slow with respect to disk rotation and large-scale with respect to epicycle radii. In this way, both Rossby vortices in planetary atmospheres and spiral structures in gaseous galactic disks can be qualitatively treated on the basis of two-dimensional fluid dynamics.

3.2 Structure Generation
by Flows with Horizontal Velocity Shear

From Figs. 2.1,2 we see that all major vortices on Jupiter are located at those latitudes where the horizontal gradient of the zonal flow velocity passes through a local maximum. This alone is enough to suggest that the vortices are associated with instabilities of zonal flows. The Rayleigh-Kuo criterion for zonal flow instability in the constant thickness approximation (this criterion is improved in Sect. 7.6 to account for the deformation of free surface) has the form [3.4, 5]

$$u''(y) - \beta(y) = 0 \tag{3.1}$$

where u'' is the second derivative of horizontal flow velocity with respect to the meridional coordinate and $\beta(y)$ is defined by (2.12). According to [3.6,7], the condition (3.1) is satisfied within the system of Jovian zonal flows exactly at those latitudes where the long-lived vortices are observed (Fig. 3.1). A similar situation is found on Saturn: the zonal flow velocity is about four times as large there as on Jupiter and the Rhines length (2.19) is twice as large; therefore the value of u'' is the same and the relationship between u'' and β, which determines the criterion (3.1), is approximately the same as on Jupiter [3.8–10].

A good example of a large Jovian vortex generated by zonal flows is the case of JLRS. This vortex is not permanent: it disappears after about five years and then reappears again within two years.

As indicated in Sect. 2.4, the instability of sheared flows is also a fundamental ingredient in the hydrodynamic concept of galactic spiral structure genesis. Thus there are grounds to assume that all the natural structures in question, both

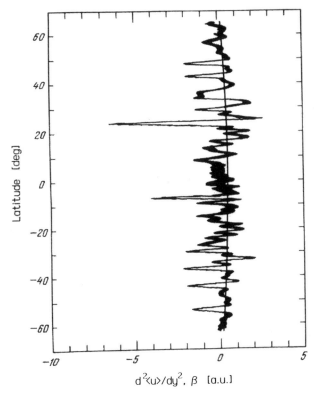

Fig. 3.1. Second derivative of zonal flow velocity averaged along the parallel $d^2\langle u\rangle/dy^2$ in Jovian atmosphere *versus* planetary latitude (*line width* indicates measurement error), in relative units, and meridional profile of β (*smooth curve*), in the same units; positive direction of the meridional coordinate y is to the north [3.7]

planetary and galactic, are produced by a common mechanism, namely the hydrodynamic instability of flows with a horizontal velocity shear, that is, with a velocity shear in the plane normal to the local angular velocity vector of the system.

3.3 Horizontal Dimensions Exceeding the Characteristic Rossby Radius

As reiterated above (Sect. 2.2), the dimensions of the long-lived vortices on Jupiter and Saturn exceed the baroclinic Rossby radius r_i, while the largest of these (the JGRS) is even bigger than the Rossby-Obukhov radius $r_R = c_0/2\Omega_z$ where $\Omega_z = f/2 = \Omega\cos\alpha$ is the projection of the angular velocity vector of system rotation onto the local vertical. In galaxies there also exists a physical counterpart of r_R, the so-called epicyclic radius $r_e = c_s/2\Omega$. Its value is readily obtained when the speed of sound c_s is known and the characteristic angular velocity Ω is determined from a plot of the kind shown in Fig. 2.13. For example, $r_e \approx 0.1$ kiloparsec for a typical value of $c_s = 10$ km/s [3.11]. This radius is quite small compared to the disk radius and to the characteristic spiral arm length which are both of the order of 10 kiloparsec, or about $10^2 r_e$.

We know now that the natural phenomena of interest to us – large-scale Rossby vortices and galactic spiral structures – have important features in common:

(1) quasi-two-dimensionality,

(2) presumably, a common mechanism of generation by flows with velocity shear, and

(3) large horizontal dimensions with respect to the same scale parameter, the Rossby radius.

It must be added that the characteristic thickness of these astrophysical objects is small relative to their horizontal size. These fundamental similarities make it possible to use a common approach to the laboratory simulation of these natural phenomena. This approach is based on the study of the dynamics of rotating shallow water.

4. Physical Prerequisites of the Laboratory Simulation of Large-Scale Rossby Vortices and Galactic Spiral Structures

The simulation of astrophysical phenomena, described in this book, is not based on their outward resemblance to the structures observed in the experiments but rather on the generality of the physical laws governing the dynamics of both laboratory and real-life phenomena, despite the immense difference in their spatial and temporal scales. To illustrate this generality, we start with a very instructive example.

4.1 The Analogy Between Two-Dimensional Gas Dynamics and the Dynamics of Shallow Water

4.1.1 Theory

In an incompressible medium a fluid flow with a jump (discontinuity) in its velocity profile is always unstable: this is the well-known Kelvin-Helmholtz (K-H) instability [4.1–4]. If, however, the medium is compressible, then a flow with a discontinuity in its (tangential) velocity may become stable, as shown by Landau [4.5]. This would depend on two prerequisites: perturbations must be two-dimensional so that their wave vector has only two components, one parallel to flow velocity and the other normal to discontinuity plane; and the velocity jump across the discontinuity $2u$ must be substantially supersonic

$$2u \geq 2\sqrt{2}c_s \ . \tag{4.1}$$

A similar theoretical result has been obtained for a tangential velocity jump in shallow water, that is, in a free-surface layer of incompressible fluid in the gravitational field, with layer thickness H_0 small compared to any relevant horizontal dimensions [4.6]. By definition, in shallow water all flow velocity perturbations are two-dimensional and lie in the horizontal plane [4.1]. Therefore, if there exists a velocity jump between adjacent regions, then it necessarily satisfies the first of the above-mentioned Landau conditions. Then, according to [4.6], the tangential velocity jump will be stable if

$$2u \geq 2\sqrt{2}(gH_0)^{1/2} \ . \tag{4.2}$$

Obviously, the criterion (4.2) is a physical equivalent of (4.1). As a matter of fact, there is an analogy between two-dimensional gas dynamics and the dynamics of shallow water: the compressions and rarefactions of a compressible gas correspond to the elevations and depressions on the surface of an incompressible fluid, while the speed of sound in the gas corresponds to the characteristic velocity of waves in shallow water [4.1]

$$c_s \rightarrow c_0 = (gH_0)^{1/2} \; . \tag{4.3}$$

(If the water is rotating, the value of g must be replaced by the net acceleration g^* due to the joint action of gravity and the centrifugal force caused by overall system rotation.) In this sense, the motion of shallow water with velocity $u >$ c_0 can be termed "supersonic". To put it another way: in shallow water, as in any incompressible fluid, the three-dimensional velocity divergence is zero, whereas its two-dimensional divergence may be nonzero owing to elevations and depressions on the free surface of the liquid. This can be seen from the continuity equation for shallow water

$$\partial H / \partial t + \mathrm{div}(H\boldsymbol{v}) = 0 \; . \tag{4.4}$$

Since $H \neq$ const for perturbations in free-surface shallow water, it follows that $\mathrm{div}\boldsymbol{v} \neq 0$, as opposed to the case of an incompressible medium of density ϱ with no free surface, where $\partial \varrho / \partial t + \mathrm{div}(\varrho \boldsymbol{v}) = 0$ and $\varrho =$ const imply $\mathrm{div}\boldsymbol{v} = 0$.

Relations (4.1–3) express the mentioned physical analogy between two-dimensional fluid dynamics and the dynamics of shallow water which can be verified experimentally. In such experiments we used an arrangement with differentially rotated shallow water (Sect. 6.3) whose geometry and other parameters were specially designed for the purpose of simulating all the relevant astrophysical phenomena [4.7, 8]. Attention must be drawn to the fact that the rotation of shallow water enters as a new physical circumstance which is not accounted for in the theory leading to (4.1, 2). Here one must distinguish between two cases. In the first case, when, of the two shallow water areas between which there is a velocity jump, the one closer to the axis is rotating faster (the "anticyclonic jump"), the rotation may *qualitatively* change the stability criteria. In this case, as will be shown below (Sect. 8.3), in addition to the K–H instability there arises the so-called centrifugal instability which has a different mechanism and therefore does not comply with the stabilization conditions (4.1, 2). (The centrifugal instability is in a sense similar to the equilibrium instability of a heavier liquid on top of a lighter one.) In this chapter we shall refrain from discussing the anticyclonic jump and consider instead the second case in which the faster rotating shallow water area is farther from the axis than the one rotating slower (the "cyclonic jump"). In this case, the rotation of shallow water results only in an additional small quantitative correction to the condition (4.2) modified by the replacement $g \rightarrow g^*$ mentioned above. This correction is of no fundamental importance since what we are concerned with is the qualitative analogy between the dynamics of shallow water and the two-dimensional gas dynamics of astrophysical phenomena.

4.1.2 Experiment

The experimental configuration is shown in Fig. 4.1. Since the free surface of a rotating liquid assumes the shape of a paraboloid, the vessel bottom is also made approximately paraboloidal in order to obtain a water layer of nearly constant thickness. To produce flows in the liquid, circular grooves were made in the vessel bottom. The grooves trace the "parallels" of the vessel and are some distance δ apart along the meridian. There are rings seated in the grooves flush with the bottom; they can be rotated at variable speeds. The outer ring is rotated faster than the vessel and the inner one is rotated slower. The angular velocities of the rings with respect to the reference frame of the vessel are equal in magnitude. The rotating rings entrain the overlying liquid layers and thus create counterflows in the frame of the rotating vessel. The liquid layer is a few millimeters thick. The counterflows thus produced have a cyclonic velocity shear: the vorticity vector on the border between the flows is parallel to the local vertical component of the system angular velocity. Ring separation is made as small as possible in these experiments ($\delta = 1$ mm) so as to obtain conditions close to a tangential jump in the velocity profile.

The experiments have shown that if the magnitude u of the flow velocities relative to the rotating vessel exceeds a certain low threshold u_{c1}, the flows tend to develop the K-H instability, resulting in a chain of cyclonic vortices like that shown in Fig. 4.2.

If the velocity of the flows is so large that it exceeds a certain relatively high threshold u_{c2}, then the flow system remains practically stable. The threshold velocity u_{c2} is the lower boundary of this stability region: as the flow velocity is being lowered we observe at $u = u_{c2}$ the transition from the stable regime to the unstable one in which a chain of cyclones is generated (Fig. 4.2). The value of u_{c2} depends on liquid layer thickness H_0, as shown in Fig. 4.3. It is easy to see that the condition of counterflow stability in the experiment, $u > u_{c2}$, is very close to the theoretical condition (4.2) for the breakdown of the tangential velocity jump instability in the "supersonic" regime. As regards the quantitative aspect of these experiments, the measurements indicate that the ratio $u_{c2}/(g^*H_0)^{1/2}$ is somewhat smaller than the theoretical value of $\sqrt{2}$, see Fig. 4.3. However, this discrepancy, attributed to the stabilizing effect of rotation in the case of a cyclonic velocity jump, matters little in the qualitative respect.

Thus, the results illustrated in Figs. 4.2, 3 can be regarded as an experimental proof of the physical analogy between two-dimensional gas dynamics and the dynamics of free-surface incompressible shallow water.

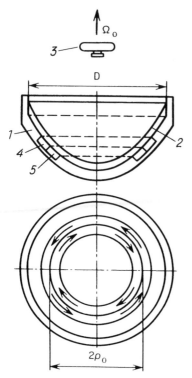

Fig. 4.1. Experimental configuration for studying the instability of tangential velocity jump in free-surface shallow water: (**1**) vessel with nearly parabolic bottom; (**2**) surface of the liquid which spreads evenly over the bottom as the vessel rotates; (**3**) camera; (**4, 5**) bottom rings for producing zonal counterflows (*arrows* in the plane view indicate flow directions with respect to the vessel for the case of a cyclonic velocity jump). The vessel and the camera rotate at angular rate $\Omega_0 = 10.85$ rad/s; $\varrho_0 = 10$ cm is the radius of the tangential velocity jump; $D = 28$ cm

4.2 Principal Similarity Parameters in Nature and Experiment

Let us now summarize the physical conditions which must be satisfied for a successful simulation of the relevant quasi-two-dimensional natural phenomena in laboratory experiments with rotating shallow water. These conditions impose certain restrictions on motion characteristics and experimental arrangements. In discussing these restrictions, we shall use such similarity parameters as the Rossby-Kiebel (Ro= v/lf), Froude (F= l^2/r_R^2), Mach (M= $u/(g^*H_0)^{1/2}$), Reynolds (Re= lv/ν), and Eckman (E= l_E^2/H_0^2) numbers, where l is the characteristic size of the structure, f is the Coriolis parameter, ν is the kinematic viscosity, and

$$l_E = (2\nu/f)^{1/2} \tag{4.5}$$

is the thickness of the so-called Eckman bottom layer. The estimate (4.5) for l_E and for the characteristic time of Eckman damping is easy to obtain from the following simple considerations. Equating the Coriolis force $f \cdot v$ to the force of friction between the liquid and the bottom $F = \nu \Delta_z v \approx \nu v / l_E^2$ where l_E is the characteristic scale of the vertical velocity gradient at which this equation can be satisfied, we obtain $l_E = (\nu/f)^{1/2}$; this is the thickness of the viscous Eckman bottom layer (4.5) we sought for. The characteristic time of motion damping in this layer due to viscosity will be $l_E^2/\nu \approx f^{-1}$. Therefore the damping time τ_E of a two-dimensional motion in a liquid layer of thickness H_0 will be larger by a factor of H_0/l_E, that is, $\tau_E \approx H_0/(\nu f)^{1/2}$. This estimate differs from (4.8) below by only a factor of $\sqrt{2}$.

The angular frequency of system rotation must be sufficiently high. This follows from the requirement that the Rossby-Obukhov radius defined by (2.6) must be much smaller than the characteristic size of the experimental installation D (its diameter), that is, $r_R \ll D/2$, to conform to the situation in the atmospheres of the giant planets, in the oceans, and in galaxies.

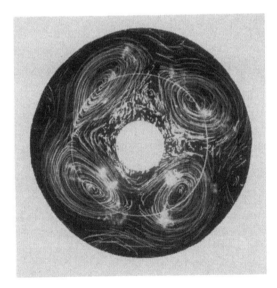

Fig. 4.2. Flow pattern in free-surface shallow water in the case of an unstable cyclonic velocity jump in the system of counterflows (experimental configuration as in Fig. 4.1; the white circle passing through cyclone centers is the line of tangential velocity jump)

The thickness of the liquid layer H_0 must be small compared to the characteristic horizontal scale parameter, the Rossby-Obukhov radius: $H_0 \ll r_R$. On the other hand, layer thickness must not be too small; it must be appreciably larger than the thickness of the viscous Eckman bottom layer within which the velocity of the liquid varies strongly along the vertical, i.e. $H_0 \gg l_E$. This implies

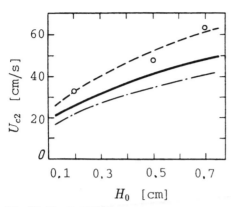

Fig. 4.3. Threshold value u_{c2} at which the tangential velocity jump becomes unstable as the flow velocity is decreased, as a function of liquid layer thickness H_0: *circles* and *dashed curve* show the velocity of rings driving the liquid above them; *dots and dashes* show liquid velocity at its surface; *solid curve* is the theoretical threshold flow velocity $u = (2g^* H_0)^{1/2}$

$$E \ll 1 . \tag{4.6}$$

If this condition is satisfied, viscosity does not affect the main features of the relevant phenomena and liquid motion can be considered to be two-dimensional throughout the entire layer.

The condition (4.6) is qualitatively equivalent to the condition of slow dissipation

$$f\tau_E \gg 1 \tag{4.7}$$

where

$$\tau_E = \sqrt{2}H_0/(\nu f)^{1/2} \tag{4.8}$$

is the characteristic "Eckman" viscous damping time of the velocity of a plane-parallel flow in an incompressible rotating liquid with no free surface (the viscous damping time of Rossby vortices in a free-surface shallow water is much longer – see Sect. 5.6 for more details).

The laboratory vortices simulating the long-lived vortices of planetary atmospheres and oceans have to be large-scale – that is, their horizontal dimensions must exceed the Rossby-Obukhov radius. In other words, the Froude number must be greater than unity: $F > 1$.

The vortex amplitude must not be small, since in natural (atmospheric and oceanic) Rossby vortices the characteristic fluid velocity is considerably higher than the vortex drift velocity $V_{dr} \approx V_R$. Therefore, according to (2.11),

$$g(\delta H)_0/af > V_{dr} , \quad \text{or} \quad h_0 \equiv (\delta H)_0/H_0 > a/R$$

where R is the system curvature radius.

The vortices under consideration which simulate vortices in planetary atmospheres and oceans must be within the Rossby regime: $Ro \ll 1$. This imposes restrictions on the amplitude of layer thickness deviation $(\delta H)_0$ and hence on the amplitude of fluid velocity within the vortices (angular ω_0 or linear v_0)

$$Ro = \omega_0/f \approx v_0/af \approx g(\delta H)_0/(af)^2 \ll 1$$

where we have used the equation of geostrophic equilibrium (2.11).

In the laboratory simulation of vortices generated by atmospheric flows, the counterflows' velocity profile should comply with the actual planetary conditions: its width must be adjustable from values below r_R, corresponding, for example, to the generation of the cyclonic Brown Ovals of Jupiter, to values above r_R, corresponding to the generation of the anticyclonic JGRS (of course, the sign of the counterflow vorticity should be adjustable as well).

The velocity profile of the flows must comply with the Rayleigh-Kuo instability condition (3.1).

In the laboratory simulation of galactic spiral structures, the velocity profile must include a jump between the faster rotating central region of the liquid (the "core") and the slower outer region (the "periphery").[1] The peripheral angular velocity must correspond to the above condition of a sufficiently large Froude number $F > 1$.

The magnitude of the velocity jump in sheared shallow water flows must correspond to: (a) the subsonic regime in planetary atmospheres, and (b) the supersonic regime in galaxies. In the latter case, the magnitude must comply with $u \gg (g^*H_0)^{1/2}$, that is, the Mach number must be $M \gg 1$.

The Reynolds number must be large enough and variable within a sufficiently wide range: $Re = au/\nu \gg 1$.

Our experimental installations stand up to all of these requirements, since the parameters listed above are, as a rule, variable within the following ranges: $F = 2$–9; $M = 0.3$–20; $Re = 5 \cdot 10^2$–$5 \cdot 10^3$; $E = 10^{-4}$–$2 \cdot 10^{-1}$; $h_0 = 0.2$–1.

---

[1] The idea that the hydrodynamic mechanism which generates galactic spiral structures could be simulated by means of laboratory experiments with differentially rotating shallow water was first suggested by A.M. Fridman in 1983.

5. Physical Basis for the Experimental Investigation of Rossby Solitons and Laboratory Simulation of Drift Vortices and Solitons in Magnetized Plasma

Structures similar to Rossby vortices in rapidly rotating hydrodynamic media can also exist in strongly magnetized plasma. This follows from the analogy between the motion of a neutral particle in a noninertial (rotating) frame under the influence of the Coriolis force and the motion of a charged particle under the influence of the Lorentz force. In order to understand the physics underlying this analogy, we begin with a few simple examples.

5.1 Two-Dimensional Motion of Charged Particles in Magnetized Plasma and Particles in Rotating Shallow Water

Let a gas particle of mass M move freely (with no collisions) at a velocity c_0 normal to the local vector of system angular velocity $\Omega_z = \Omega \cos \alpha$. Then, due to the Coriolis force, the horizontal projection of particle motion will describe a circle. The characteristics of particle revolution are determined by $M c_0^2/r = M f c_0$. That is, the radius of the circle is

$$r = \frac{c_0}{f} = \frac{(g^* H_0)^{1/2}}{f} = r_R$$

and the angular velocity of particle revolution equals f. On the other hand, a positive plasma ion of mass M and charge e whose velocity $c_s = (T_e/M)^{1/2}$ is normal to the magnetic field B moves in a Larmor orbit of radius

$$r_L = \frac{c_s}{\omega_B} = \frac{(T_e/M)^{1/2}}{\omega_B}$$

at an angular frequency equal to the ion Larmor frequency $\omega_B = eB/Mc$ where T_e is the plasma electron temperature, B is the magnetic field strength and c is the light velocity. This example alone illustrates the analogy between the motions in question.

Now consider the fact that in the case of free-surface shallow water, according to a result known from fluid mechanics [5.1], the role of the volume density of the medium (plasma) n_0 is played by the surface density proportional to H_0 or to $H_\eta = H_0/\cos\alpha$, the layer thickness measured along the axis η directed against the gravity vector g and coinciding with the system's rotation axis. We have then a straightforward correspondence between the parameters describing the dynamics of shallow water and magnetized plasma, see Table 5.1.

Table 5.1. Correspondence between the dynamic parameters of shallow water and magnetized plasma

Shallow water	Plasma
r_R	r_L
f	ω_B
$(gH_\eta)^{1/2}$, $(g^*H_0)^{1/2}$	$(T_e/M)^{1/2}$
H_η, H_0	n_0

If the ion orbit parameters change in the direction normal to the magnetic field B, its path becomes a trochoid rather than a circle, that is, the ion drifts in the direction normal both to B and to the direction in which the parameters change. This drift is similar to the drift of a Rossby vortex due to spatial nonuniformity of f (or H_0).

5.2 General Nonlinear Equation for Drift Motion

On the basis of what has been said in the previous section, we can introduce the notion of a Rossby regime for plasmas – meaning a slow motion occurring at a characteristic frequency $\omega < \omega_B$ in the plane perpendicular to the magnetic field. In plasma physics such motions are called drifts. Usually, we have in mind a "magnetized" collisionless plasma, where, in addition to the inequality $\omega < \omega_B$, two other conditions are also satisfied. The first is the "magnetization" condition $l > r_L$ where l is the characteristic size of the structures under consideration across the magnetic field and r_L is the above-mentioned ion Larmor radius at the electron temperature. The second is the "non-collision" condition $\omega_B\tau_i \gg 1$ where τ_i is the characteristic time of ion collisions. A parallel to the case of rotating shallow water can be easily seen here, too: the first of the two conditions is similar to the requirement for shallow water structures to be large-scale relative to the Rossby-Obukhov radius and the second formally corresponds to the condition (4.7) demanding the friction of liquid against the bottom to be weak.

Now, going over to collective drift motions of particles, we must take into account the fact that in a quasi-neutral plasma a slow perturbation δn of the electron density n_0 is uniquely related to the perturbation φ of the equilibrium plasma potential: the Boltzmann distribution implies $e\varphi/T_e = \delta n/n_0$. According to the above discussion and Table 5.1, the dimensionless perturbation $h = \delta H/H_0$ in the thickness of free-surface shallow water plays the same role in the motions occurring in the Rossby regime as does the quantity $e\varphi/T_e$ (or $\delta n/n_0$) in plasma.

In the simultaneous discussion of drift motions in the two media, shallow water and magnetized plasma, one should naturally adopt the same approximations. One of the most important approximations in geophysical hydrodynamics, upon which, in particular, our entire treatment of the theory of Rossby waves and vortices is based, is the "β-plane" approximation mentioned already in Sect. 2.1. It plays a very productive role, the essential idea being that the spherical shape of the planet is disregarded within a certain vicinity of the point under consideration and the unperturbed surface of the atmosphere (or ocean) is assumed to be flat, coinciding with the plane tangent to the actual unperturbed surface. All analysis is then carried out in this plane under the assumption that the value of β is constant (explaining the name "β-plane"). In other words, the dependence of the Coriolis parameter on the meridional coordinate y is assumed to be linear, according to (2.13). Deviations from this approximation can be taken into account in the form of corrections to the basic solution.

In the theory of plasma drift waves, the conditions of the β-plane approximation correspond to such a situation when, within an axially symmetric plasma column oriented along the magnetic field, plasma density falls off exponentially in the transverse direction (radial coordinate y), i.e. the characteristic scale of the plasma density gradient is constant: $|n_0/(\partial n_0/\partial y)| = R = \text{const.}$[1] In this case, the maximum velocity of linear drift waves – the so-called drift velocity V^* – does not depend on y and is the counterpart of the Rossby velocity V_R (however, in the more general case which we shall have to take into account later, both velocities may depend on y).

Another approximation used in geophysical hydrodynamics is the traditional quasi-geostrophic approximation, valid for motions which are small-scale relative to the Rossby-Obukhov radius. It is convenient, in particular, for the treatment of barotropic (two-dimensional) motions in the terrestrial atmosphere where r_R certainly exceeds the dimensions of natural structures. Within this approximation, the nonlinear equation for Rossby waves (vortices) becomes the well-known Charney-Obukhov equation [5.2, 3]

$$(\Delta h - h)_t + V_R h_x + J(h, \Delta h) = 0 , \tag{5.1}$$

see Supplement S5.1. Here $J(h, \Delta h) = h_x(\Delta h)_y - h_y(\Delta h)_x$ is the Jacobian; the subscripts denote differentiation with regard to the coordinates x (along the "parallel"), y (along the "meridian") and time t; the equation is written in terms of dimensionless units, the dimensioned units x, y, t and $V_R = \beta r_R^2$ (see (2.14)) having been transformed according to the relations: $x \to r_R x$, $y \to r_R y$, $f_0 t \to t$, $V_R \to (g^* H_0)^{1/2} V_R$.

Equation (5.1) corresponds to motions in the β-plane, that is, in the plane spanned by the x and y axes, and describes a structure of finite dimensions along

---

[1] Here we adopt the model assumption that the characteristic scales of electron temperature and density gradients are the same (equal to R) and the signs of the two gradients are opposite. See Supplement S5.4 for more details.

these axes. The values of the parameters are those at the center of the structure which is assumed to be at $y = 0$.

Equation (5.1) is not a newcomer in the toolkit of modern geophysical hydrodynamics and may also be found in tutorials like [5.4, 5]. Among its solutions are the dipolar Rossby solitons discussed below, very popular objects in modern hydrodynamics [5.6].

There is a quite similar equation in plasma theory, describing, in the framework of an approximation similar to the β-plane approximation, drift motions on a scale small compared to the characteristic ion Larmor radius r_L. It is called the Hasegawa-Mima equation [5.7–9] and has the same form as (5.1), the only differences being that the equation is in terms of $\delta n/n_0$ (or $e\varphi/T_e$), and the Rossby velocity V_R is replaced by its counterpart, the drift velocity V^*. Naturally, the Hasegawa-Mima equation also has dipolar solitonic (in this case, plasma drift) solutions [5.10].

Thus, within the approximations indicated above, the analogy between drift motion equations in geophysical hydrodynamics and in plasma physics becomes more complete.

Now, going back to hydrodynamics, we will surpass the limitations of the geostrophic approximation and also make a correction "enhancing" the β-plane approximation. One more nonlinear term, accounting for the finite perturbation amplitude, is then added to (5.1) and the Rossby velocity becomes a function of the meridional coordinate y [5.11–25] (see Supplement S5.1):

$$(\Delta h - h)_t + V_R(y)h_x + V_R h h_x + J(h, \Delta h) = 0 . \tag{5.2}$$

For the Rossby velocity, according to (2.13), we have

$$V_R(y) = \beta r_R^2(y) = \beta g H_0 (f_0 + \beta y)^{-2} ,$$

and since $\beta \approx f_0/R$ where R is the meridional curvature radius,

$$V_R(y) \approx V_R(1 - 2R^{-1}y) \tag{5.3}$$

where $V_R = V_R(0)$. With (5.3) taken into account, (5.2) takes the form

$$(\Delta h - h)_t + V_R h_x + V_R h h_x + J(h, \Delta h) - 2V_R h_x y R^{-1} = 0 . \tag{5.4}$$

The fifth term in (5.4) which is due to spatial inhomogeneity of the Rossby velocity is of major importance for the types of nonlinear structures allowed by this equation. This fundamental question will be discussed below.

The new nonlinear term in (5.2), $V_R h h_x$, is the so-called scalar nonlinearity, physically analogous to the nonlinearity (discussed below) which determines the properties of another equation known for a whole century, the classical Korteweg-de Vries (KdV) equation. It describes the Russell soliton, the first soliton in the history of science, observed more than a century and a half ago [5.26]. A similar soliton, the so-called monopolar Rossby soliton, has been discovered theoretically by Flierl [5.12] on the basis of (5.2). This, too, will be discussed in detail below.

All what has been said up to this point in the present section allows us to skip the derivation of (5.2) and turn directly to the analysis. Note that the quasi-geostrophic equation (5.1) in the general case is often written in terms of the so-called stream function ψ rather than in terms of h. The derivatives of this function with respect to coordinates x and y are velocity projections onto the y and x axes, respectively. In the case of a free surface, ψ is related to h by $\psi = (g/f_0)h$, as can be readily seen from the equation of geostrophic equilibrium (2.11).

Let us now turn to the equation for drift waves in magnetized plasma ([5.7–10, 19, 20, 22–25], Supplement S5.4). Similar to (5.2), we obtain (see the last footnote above):

$$(\Delta\varphi - \varphi)_t + V^*(y)\varphi_x + V^*\varphi\varphi_x + J(\varphi, \Delta\varphi) = 0 \tag{5.5}$$

where the following transformations have been applied to the dimensioned quantities x, y, t, φ, and $V^* = \beta r_L^2$ in order to make them dimensionless: $x \to r_L x$, $y \to r_L y$, $\omega_B t \to t$, $e\varphi/T_e = \delta n/n_0 \to \varphi$, and $V^* \to (T_e/M)^{1/2}V^*$, where

$$\beta = \omega_B \frac{\partial(\ln n_0)}{\partial y} = \frac{\omega_B}{n_0}\frac{\partial n_0}{\partial y} , \tag{5.6 a}$$

in accordance with the readily obtainable result that

$$\beta = f_0 \frac{\partial(\ln H_\eta)}{\partial y} = \frac{f_0}{H_\eta}\frac{\partial H_\eta}{\partial y} \tag{5.6 b}$$

for shallow water. Similar to the Rossby velocity equation (5.3), in this case we have for the drift velocity

$$V^*(y) = \beta r_L^2(y) = V^*[1 + T_e^{-1}(\partial T_e/\partial y)y] = V^*(1 - R^{-1}y)$$

where $V^* = V^*(0)$. Now (5.5) becomes

$$(\Delta\varphi - \varphi)_t + V^*\varphi_x + V^*\varphi\varphi_x + J(\varphi, \Delta\varphi) - V^*R^{-1}\varphi_x y = 0 . \tag{5.7}$$

It is evident that (5.4) and (5.7) are virtually the same. This fact, as well as the similarity of the two equations (5.6), illustrates the remarkable analogy between Rossby waves in shallow water and drift waves in plasma. Keeping this analogy in mind, we shall henceforth assume the qualitative conclusions drawn from the equation for Rossby waves to be applicable also in the case of plasma drift waves.

It should be noted that the relation $H_0 = H_\eta \cos\alpha$ used above in the derivation of (5.6b) is valid for planets only if the value of $\cos\alpha$ is not too small – that is, for not very low latitudes. This, however, does not reduce its generality, since the entire theory of Rossby waves is based on assumptions which hold only for "intermediate" latitudes, losing their validity, in particular, near the equator where $\cos\alpha \to 0$ [5.4, 27].

The approximation used here as well as in the theory of Rossby waves is compatible with the experimental configurations we have employed for studying Rossby vortices (Sect. 6.1). The relation $H_0 = H_\eta \cos \alpha$ holds for all the available values of α in these configurations. A more precise expression for β, free from this restriction, follows from (5.11) below. At present, we stick to (5.6b) in order to emphasize the symmetry between the expressions for β in shallow water and in plasma.

5.3 Linear Rossby Waves and Plasma Drift Waves

Equations (5.4) and (5.7) are essentially nonlinear. Now consider the small amplitude approximation in which they can be linearized by going over to the linear regime: $\partial/\partial t \to -i\omega$, $\partial/\partial x \to ik_x$, $\Delta \to -k^2$; $k^2 = k_x^2 + k_y^2$. Here k_x and k_y are the wavenumbers corresponding to the directions along the parallel (positive "eastwards", in the sense of system rotation) and along the meridian (positive "northwards", towards the system rotation axis). Neglecting the spatial nonuniformity of the velocities $V_R(y)$ and $V^*(y)$, from (5.4) and (5.7) we obtain the dispersion equation which in terms of dimensionless quantities has the form

$$\omega = -\beta \frac{k_x}{k_x^2 + k_y^2 + 1/r_0^2} \tag{5.8}$$

where r_0 ($= r_R$ or r_L) is the characteristic dispersion scale. This equation describes linear wave motions in the Rossby regime [5.4, 27]. They are referred to as Rossby waves in geophysical hydrodynamics and as drift waves in plasma physics. The minus sign in (5.8) implies that at $\beta > 0$ (this corresponds, for instance, to an atmosphere of uniform thickness), the phase velocity of the Rossby waves has the opposite sign as the overall system rotation. On a planet, it is directed westwards. Since the magnitudes of r_R on planets are large (thousands of kilometers in planetary atmospheres), Rossby waves are the longest waves in planetary atmospheres and in oceans. They are called inertial-gravitational waves. Their observation on the Earth has been described in [5.28–30], and on one of the giant planets, Saturn, in [5.31–33] (see Fig. 5.1).

The theory of Rossby waves is usually studied within the β-plane approximation. In particular, this approximation assumes that long waves ($\lambda = 2\pi/k_x \gg r_R$) satisfy the condition $\lambda \ll R$ where R is the system curvature radius (the planet radius). This condition is satisfied on the giant planets but not on the Earth.

Equation (5.8) also describes drift waves in magnetized plasma. We are concerned here with the "hot electron" plasma where the temperature of the electrons is much higher than that of the ions. For the sake of simplicity we also assume that the longitudinal (that is, directed along the magnetic field) wavenumber $k_\parallel$ is much smaller than the transverse wavenumber $k_\perp$, so that we may set $k_\parallel = 0$ and $k \approx k_\perp$. The inclusion of $k_\parallel$ would not yield any substantial changes [5.20].

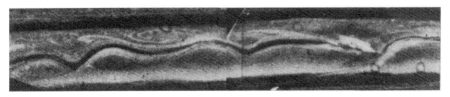

Fig. 5.1. A Rossby wave in the atmosphere of Saturn, described by dispersion equation (5.8); wave length along the parallel (horizontal direction in the figure) is about 5700 km. From [5.32]

Despite the fact that drift waves are prominently stretched along the magnetic field while Rossby waves propagate in a thin layer, the *transverse* motions which occur normal to B in plasma and normal to Ω_z in shallow water are quite similar in both cases. This corresponds well to the CGL theorem [5.34], see Sect. 3.1. We see from (5.6, 8) that the existence of drift waves and Rossby waves depends on the presence of meridional gradients of media parameters ($\beta \neq 0$). For this reason, they may be called gradient waves. Both kinds of waves travel in a direction normal to two vectors: the system angular velocity vector and the (meridional) gradient of shallow water thickness H_η – in the case of Rossby waves, the magnetic field vector and the (radial) gradient of plasma density – in the case of drift waves.

According to (5.8), the largest phase velocity V_0 of these linear waves is

$$V_0 = V_R = \beta r_R^2 = \frac{g^* \cos \alpha}{f} \frac{\partial H_\eta}{\partial y} \tag{5.9}$$

in the Rossby wave case (the so-called Rossby velocity), and

$$V_0 = V^* = \beta r_L^2 = \frac{c_s^2/n_0}{\omega_B} \frac{\partial n_0}{\partial y} \tag{5.10}$$

in the case of drift waves in plasma (the so-called drift velocity). The similarity between (5.9, 10) is in line with the analogy discussed in Sect. 5.1. Just like (5.6b), (5.9) has been obtained by substituting $H_0 = H_\eta \cos \alpha$ into the more precise expression

$$V_R = \beta r_R^2 = g^* \frac{\partial (H_0/f)}{\partial y} = -\frac{g^*}{R} \frac{\partial (H_0/f)}{\partial \alpha} . \tag{5.11}$$

When $H_0 = \text{const}$,

$$V_R = r_R^2 \frac{\partial f}{\partial y} = -\frac{r_R^2}{R} \frac{\partial f}{\partial \alpha}, \quad \beta = \frac{\partial f}{\partial y} . \tag{5.12}$$

This special case, equivalent to (2.12–14), normally corresponds to conditions in planetary atmospheres: the Rossby wave is caused by the meridional gradient of the Coriolis parameter. In general, the Rossby wave depends on the gradients of both f and H_0.

The dispersion equation (5.8) plays an important role in our discussion. We shall therefore outline two independent ways of obtaining it. First, one could follow the conventional procedure. From the Euler equation (2.1) and the continuity equation (4.4) for free-surface shallow water we can get a solution which, when expanded in powers of the Rossby number (2.3) and then linearized, yields (5.8).

Alternatively, one could use the Rossby-Ertel theorem stating that a potential vortex is "frozen-in" into fluid particles in the absence of dissipation [5.4, 27]. For shallow water, the theorem has the form

$$\frac{d}{dt}\frac{\operatorname{curl} v + f}{H} = 0 .$$
(5.13)

The first term in the numerator is the local velocity curl, f is the vector Coriolis parameter, and $H = H_0 + \delta H$. The quantity in (5.13) which is to be differentiated is known as the potential vortex, thus (5.13) is the equation of potential vortex conservation. In the special case of H = const, (5.13) may be regarded as the geostrophic generalization of the classical theorems of Helmholtz and Kelvin stating the conservation of the total curl (see, for instance, [5.4, 5]).

It should be noted that there is a counterpart of the Rossby-Ertel theorem in the case of plasma drift waves [5.22–25], namely

$$\frac{d}{dt}\frac{\operatorname{curl} v + \omega_B}{n} = 0 .$$
(5.14)

The relations (5.13, 14) characterize the general analogy between Rossby waves in shallow water and drift waves in plasma. This analogy is summarized in Table 5.2.

Table 5.2. Analogy between Rossby waves on rotating free-surface shallow water (laboratory, atmosphere, ocean) and drift waves in magnetized plasma. [Here we adopt the model assumption that the characteristic scales of electron temperature and density gradients are the same (equal to R) and the signs of the two gradients are opposite. See Supplement S5.4 for more details]

Shallow water	Plasma
Velocity of gravity waves $c_0 = (gH_\eta)^{1/2}$	Velocity of ion sound $c_s = (T_e/M)^{1/2}$
Thickness of shallow water layer H_η $h = (\delta H)_\eta/H_\eta$	Volume density of plasma n_0 $e\varphi/T_e = \delta n/n_0$
Coriolis parameter f	Ion Larmor frequency ω_B
Rossby-Obukhov radius $r_R = (gH_\eta)^{1/2}/f$ $\beta = (f/H_\eta)\partial H_\eta/\partial y = f/R$	Ion Larmor radius $r_L = (T_e/M)^{1/2}/\omega_B$ $\beta = (\omega_B/n_0)\partial n_0/\partial y = \omega_B/R$
Rossby velocity $V_R = \beta r_R^2 = gH_\eta/fR$	Drift velocity $V^* = \beta r_L^2 = c_s^2/\omega_B R$
Rossby-Ertel theorem $\frac{d}{dt}\frac{\operatorname{curl} v+f}{H_\eta} = 0$	Analog of Rossby-Ertel theorem $\frac{d}{dt}\frac{\operatorname{curl} v+\omega_B}{n} = 0$
Charney-Obukhov equation $(\Delta h - h)_t + V_R h_x + J(h, \Delta h) = 0$	Hasegawa-Mima equation $(\Delta\varphi - \varphi)_t + V^*\varphi_x + J(\varphi, \Delta\varphi) = 0$
General equation for Rossby waves $(\Delta h - h)_t + V_R h_x + V_R h h_x$ $+J(h, \Delta h) - 2V_R R^{-1}h_{xy} = 0$	General equation for drift waves $(\Delta\varphi - \varphi)_t + V^*\varphi_x + V^*\varphi\varphi_x$ $+J(\varphi, \Delta\varphi) - V^*R^{-1}\varphi_{xy} = 0$

In our case of two-dimensional motions we can introduce a stream function ψ whose derivatives $(-\partial\psi/\partial y, \partial\psi/\partial x)$ are the latitudinal and longitudinal velocity components. In the regime of geostrophic equilibrium (2.11) we have $\delta H = (f/g^*)\psi$, or $h = (f/g^*H_0)\psi$. Expanding $(1 + \delta H/H_0)^{-1}$ and retaining terms up to the second order, we can use the Rossby-Ertel theorem to obtain the differential equation (5.2). If, on the other hand, only the first-order terms are retained and then linearized, we arrive at the dispersion relation (5.8).

Using (5.13), the following simple interpretation can be given for the mechanism of Rossby wave generation. Consider a planet with a uniform thickness of its atmosphere ($H_0 = $ const) and its Coriolis parameter increasing northwards. Suppose that an atmospheric perturbation occurs at a certain latitude, resulting in some particles being displaced to the north and some to the south. This displacement is shown schematically in Fig. 5.2 by the sine-shaped solid curve. According to (5.13), a northward displacement (in the direction of increasing f) will decrease curl v. In particular, if curl $v = 0$ before the perturbation, an anticyclonic vorticity will appear: since the planet, viewed from the north, is rotating counter-clockwise, the particles which have been displaced northwards (upwards in our scheme) must receive a clockwise spin. For example, in the vicinity of point A, particles on the left will move northwards and those on the right will move southwards (as shown by arrows). A similar displacement, but with the opposite vorticity, will occur in the neighborhood of point B. (In the vicinity of point O the medium is not displaced in the meridional direction). As a result of all these motions, the initial displacement pattern will be shifted as shown by the dotted curve in Fig. 5.2 – that is, to the left. This is a Rossby wave, traveling westwards when $H_0 = $ const.

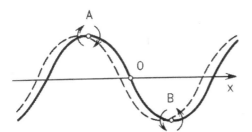

Fig. 5.2. Schematic illustration of Rossby wave formation

It can be shown that the generation mechanism of a drift wave in a plasma is, in principle, similar to that described above.

Now let us look at (5.11). One can see that the Rossby wave is due to the spatial nonuniformity of both the Coriolis parameter ($\partial f/\partial y \neq 0$) and the shallow water thickness H_0 in the meridional direction y. The fact that Rossby wave propagation is equally affected by the gradients of both f and H_0 becomes clear from the following straightforward arguments. First, from the Rossby-Ertel theorem (5.13) which defines the dispersion equation (4.8) we see that vortex dynamics depends equally on the variables $H_0^{-1}\partial H_0/\partial y$ and $-f^{-1}\partial f/\partial y$, in line with (5.11). Second, from the geostrophic equilibrium equation (2.4) it is easy to find a gradient $\partial H_0/\partial y$ that would give rise to the same drift as the gradient $\partial f/\partial y$. For example, let the gradient of f cause a drift at a velocity of $V_{dr} \approx V_R = r_R^2 \partial f/\partial y$. From (2.4) we find that the same drift rate would be produced by a gradient of H_0 such that $H_0^{-1}\partial H_0/\partial y = -f^{-1}\partial f/\partial y$. It is exactly for this reason that (5.11) incorporates the sum of these two gradients (with the appropriate signs): $V_R \propto (H_0^{-1}\partial H_0/\partial y - f^{-1}\partial f/\partial y)$.

An interesting consequence of this fact is that a "wind" with velocity V_R in the direction of Rossby wave propagation produces, due to the Coriolis force, a gradient $\partial H_0/\partial y$ that exactly counterbalances the dispersion (Sect. 7.6).

From the above discussion we see that a Rossby wave does not necessarily travel against the direction of system rotation: if there is a high enough gradient of shallow water thickness and its direction coincides with that of the Coriolis parameter gradient, then the Rossby wave phase velocity is directed *along* the direction of system rotation. This situation is easily created in the experiments described in Chapt. 9.

5.4 Linear Packet of Rossby Waves and the Time of its Dispersion Decay. Definition of a Soliton

Much attention will be paid in this book to Rossby solitons rather than just Rossby vortices. A soliton is a nonlinear wave packet which is practically immune to dispersion decay. Either it undergoes no dispersion decay at all, or the characteristic time of its dispersion decay τ_d is much greater than the dispersion spreading time τ_1 of a linear wave packet of the same size.

This definition of a soliton does not differ from the definition of a solitary wave. However, the term "solitary wave" does not suit the dual-natured objects of our attention, combining wave features with clearly pronounced vortical properties. Therefore we find it preferable to use the term "soliton" as regards these objects. The above definition of solitons is irrespective to the character of their collisions. Thus it differs from the definition (favored mainly by mathematicians) according to which only those solitary waves are called solitons that collide elastically with one another [5.26,35]. We apply the latter definition only to the "pure wave" solitons (not even to all of them); therefore we do not use it in this book because, as we shall see, "pure wave" Rossby solitons do not exist. As regards the vortical solitons, or solitonic vortices, which are the main object of our discussion, the term "solitary vortex" [5.15–17] seems quite appropriate since it combines the notion of a vortex with the implication of a solitary wave.

Thus, in order to identify convincingly a Rossby vortex as a soliton in an experiment, one must be able to show that its dispersion decay time τ_d is large enough:

$$\tau_d \gg \tau_1 . \tag{5.15}$$

Let us therefore express τ_1 as a function of linear packet parameters. For the Gulf Stream rings the problem of dispersion spreading for a linear two-dimensional (round) Rossby wave packet has been solved analytically within the β-plane approximation, neglecting viscosity [5.36]. The main goal of that study was an attempt to answer the question whether the Gulf Stream rings are long-lived (compared to linear wave packets of the same size). In accordance with the geometry of the problem it was assumed that the dispersion of Rossby waves was caused only by the curvature of the planet surface and that (2.12) was valid since $H_0 = $ const.

The results obtained in [5.36] are reproduced in Fig. 5.3 where three parameters of the packet are plotted versus its radius: the characteristic lifetime τ_1, the westward (counter to planet rotation) drift velocity V_l, and the free-run length over the lifetime s_l (initial profile of the packet is Gaussian (2.20)). The quantity τ_1 is the time required for the packet amplitude to decrease twofold due to dispersion. One can see that the time of dispersion spreading has a minimum which corresponds to the packet radius $a \approx r_R$ and equals

$$(\tau_1)_{min} \approx 8 \frac{r_R}{V_R} , \tag{5.16}$$

as indicated in Sect. 2.1. The propagation velocity of such a packet is approximately one-third the Rossby velocity V_R, and the distance it covers during its lifetime (5.16) does not exceed one and a half of the length of its diameter

$$(s_l)_{min} \approx \frac{1}{3} V_R (\tau_1)_{min} \approx 2.7 r_R . \tag{5.17}$$

The physical meaning of (5.16) can be readily illustrated with the aid of the dispersion relation (5.8), see also Fig. 5.4a. It is easy to see that for $k_y = 0$, half the period of the highest-frequency component of the packet (of all the Rossby waves), corresponding to the maximum of the function $\omega(k_x)$, is $T/2 = 2\pi r_R/V_R$ which is virtually the same as the dispersion spreading time (5.16). Clearly, a time shorter than $T/2$ is not enough for a linear wave packet to spread since the contributions of none of the Fourier components of the packet can change significantly within that time. This means that (5.16) is not an overestimate.

The nonmonotonous nature of $\tau_1(a)$ can be explained by the following argument. As the packet decreases in size, that is, as $k \to \infty$, the frequencies of the Rossby waves become very low ($\omega \to 0$), leading to a significant increase of the lifetime τ_1. On the other hand, as the packet size increases, the wave dispersion (5.8) becomes weaker. The simultaneous action of these two effects generates the minimum of the function $\tau_1(a)$.

In view of the importance of the estimate (5.16) for our further development, we shall now quote the results of a numerical computer calculation carried out

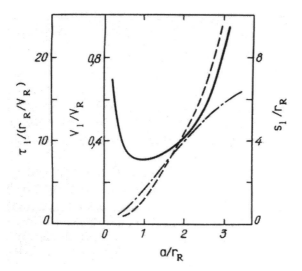

Fig. 5.3. Characteristics of a dispersive linear Rossby wave packet *versus* its radius a: lifetime τ_1, that is, the time during which packet amplitude decreases by a factor of 2 (*solid curve*); drift velocity V_l (*dots and dashes*); run length s_l (*dashed curve*). Initial packet profile $h = h_0 \exp(-r^2/a^2)$. From [5.36]

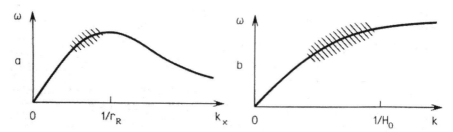

Fig. 5.4. Dispersion curves (**a**) for Rossby waves with $k_y = 0$ and (**b**) for gravity waves in a layer of nonrotating free-surface liquid: wave angular rates ω as functions of the respective wave vectors k_x and k; *hatching* shows the range where scalar KdV-type solitons can exist

by A.V. Khutoretsky, substantiating this estimate. The calculational procedure is as follows. An initially Gaussian (see (2.20)) linear packet of Rossby waves is expanded into a Fourier series of 64 harmonics with different wavenumbers and frequencies, corresponding to the dispersion equation (5.8). Each harmonic is propagated independently at its own phase velocity ω/k_x. After a predetermined time interval, the computer adds up all the harmonics and outputs the parameters of the distorted wave packet. In this way, in particular, the time $\tau_F = \tau_{\text{Fourier}}$ required to decrease by a factor of $1/e$ the velocity v_0 related by (2.11) to the amplitude of relative layer thickness deviation $h_0 = (\delta H)_0/H_0$ is determined as a function of a. At $a = r_R$ the quantity τ_F has a minimum equal to

$$(\tau_F)_{min} \approx 7 \frac{r_R}{V_R} , \qquad (5.18)$$

practically coinciding with (5.16). Generally speaking, it is this value (5.18) that should be used in comparisons with the experimental results described below. Since, however, it differs little from the earlier result (5.16), we shall also use (5.16), for "historical" reasons.

A linear packet of Rossby waves may be subject to an additional nonviscous decay, not accounted for in [5.36]. Since the Rossby velocity varies over the meridional cross-section of the packet, different parts of the packet drift at different velocities, which gradually destroys the packet. The characteristic time of this process τ_1' is accounted for by the fifth term in (5.4). From the derivation of (5.4) it is easy to see that this time is

$$\tau_1' \approx \frac{R}{2V_R} \qquad (5.19)$$

where R is the meridional curvature radius of the system ($R \approx f/\beta$) [5.15–17, 21]. Indeed, suppose the packet is moving as a whole (body) with a velocity that is a certain fraction of the Rossby velocity V_R. Since V_R is different in the "northern" and "southern" parts of the packet, the time of its shear deformation ("twisting") will be $t \approx y/\Delta V_R$ where $\Delta V_R \approx 2V_R y/R$, thus $t \approx R/2V_R$. In general, the time (5.19) is of the same order as (5.16), so the actual dispersion decay time of a Rossby wave packet is notably shorter than (5.16).

This effect must be practically absent in highly nonlinear vortices, carrying entrapped particles whose orbital velocities are much greater than the velocity of vortex drift as a body, since particles with different y coordinates will rapidly exchange their positions (see below).

5.5 Nonlinear Rossby Waves: Vortices and Solitons

Now we turn to those nonlinearities which can give rise to solitary structures (solitons) on the Rossby wave branch. Equation (5.2) involves two nonlinear terms: the third (scalar nonlinearity) and the fourth (vector nonlinearity). The scalar nonlinearity has the same form and the same physical meaning as the nonlinear term in the well-known KdV equation which describes nonlinear waves in nonrotating free-surface shallow water. In dimensioned variables it has the form [5.38, 39]

$$h_t + c_0 h_x + \frac{3}{2} c_0 h h_x + \frac{1}{6} c_0 H_0^2 h_{xxx} = 0 \qquad (5.20)$$

where the indices x and t denote partial differentiation with regard to the coordinates x and t, respectively.

Based on (5.2), one can make an order-of-magnitude estimate of the relative contributions of the scalar and vector nonlinearities [5.11]:
the scalar nonlinearity dominates if

$$a > r_R , \qquad\qquad (5.21)$$

and the vector nonlinearity dominates if

$$a < r_R \qquad\qquad (5.22)$$

where a is a characteristic size ("radius") of the structures; the radius r_R should be replaced by r_i in the baroclinic case.

The scalar nonlinearity, just as in the KdV case, is directly related to the perturbation δH of shallow water thickness H_0. The vector nonlinearity is not necessarily related to δH. In particular, it also exists in the "solid lid" case when (5.2) – coinciding in that case with (5.1) – is written in terms of the stream function ψ. Strictly speaking, the two nonlinearities can be separated only in asymptotical, modeled situations. For instance, the scalar nonlinearity vanishes when the liquid has no free surface, while the vector nonlinearity is absent under two simultaneous conditions: axial symmetry of the vortex and spatial uniformity of the Rossby velocity.

It is the regime given by (5.21) that is of interest for the physics of magnetized plasma where, by definition, $a > r_L = (T_e/M)^{1/2}/\omega_B$ for all structures. This regime is also typical for the largest vortices on the giant planets and in terrestrial oceans, while the regime given by (5.22) corresponds to large vortices in terrestrial atmosphere.

When the vector nonlinearity dominates, (5.4) becomes the Charney-Obukhov equation and allows for the existence of dipolar solitary waves, or dipolar solitons, also called "modons" (named after the abbreviation MODE standing for Mid-Ocean Dynamic Experiment, a program for the investigation of synoptic vortices in the ocean). In shallow water, they are Rossby solitons, first described by Larichev and Reznik [5.6]. In magnetized plasma, they are drift solitons [5.7, 10]. In both cases their properties are similar. A modon is an isolated cyclone-anticyclone pair of vortices. In a modon the dispersion spreading inherent to any linear wave packet is counterbalanced by the vector nonlinearity. The propagation velocity of such a soliton can belong to either of the following ranges:

a) in the same direction in which linear Rossby waves propagate (westwards on a planet), with the magnitude

$$|V_{dr}| > V_R , \qquad\qquad (5.23\,a)$$

b) in the opposite direction – with arbitrary magnitude:

$$V_{dr} > 0 . \qquad\qquad (5.23\,b)$$

These conditions mean that modon velocities are outside the velocity range of linear Rossby waves given by (5.8, 9) and shown in Fig. 5.4a, therefore modons do not waste their energy in radiating such waves through the Cherenkov mechanism. Thus (5.23) are the necessary conditions for the soliton to be stationary.

Figure 5.5 shows the prominent features of a modon. The difference from the well-known dipolar vortex in a nonrotating liquid [5.5, 40], which has a power-type velocity profile ($\propto r^{-2}$), is that the modon is more "screened-off": its velocity profile falls off exponentially at large distances from vortex center $\propto \exp(-\mu r)$. The constant μ depends on the modon drift velocity: if the latter is close to V_R then the effective modon screening radius $1/\mu$ is close to r_R. The complete expression for the screening constant is (Sect. 7.6 and [5.41])

$$\mu = \sqrt{\frac{1}{r_R^2} + \frac{\beta}{|V_{dr}|}} \ . \tag{5.24}$$

It is interesting to note that modon screening in rotating shallow water pertains even in the absence of the β-effect: according to (5.24), for $\beta = 0$ the screening radius coincides with the Rossby-Obukhov radius, $1/\mu = r_R$.

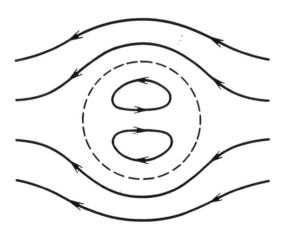

Fig. 5.5. Schematic of a dipolar Rossby soliton: streamlines in the frame which is drifting with the soliton along the parallel; *dashed curve* shows the separatrix enclosing trapped particles

A characteristic feature of the modon shown in Fig. 5.5 is the existence of inner regions with closed stationary streamlines confined within a separatrix. Particles in these regions, moving along the streamlines, are drifting with the vortex – they are trapped in it.

Finally, a more complicated kind of vector nonlinearity may arise in the case of the so-called topographic solitons. Their existence requires a special topography of the bottom [5.42]. We shall not dwell on this soliton type since our principal goal is to study the regime given by (5.21) where the scalar nonlinearity is dominating.

The scalar nonlinearity is also capable of counterbalancing dispersion. As follows from (5.2), such a balance is possible only for vortices of a definite polarity – anticyclones – where, by definition, the vorticity in the center is opposite to the local vertical component of the angular velocity of overall system rotation (see [5.11–25] for more details). The cyclone-anticyclone asymmetry of (5.2) is mainly due to the third term which accounts for the scalar nonlinearity [5.11–20]. When the dispersion and the scalar nonlinearity balance each other, we obtain a *monopolar* Rossby soliton, first described theoretically in [5.12] and later also in [5.11, 18–21].

This "scalar"-type soliton is physically similar to the well-known one-dimensional soliton, a solitary wave of elevation on the free surface of nonrotating shallow water. The latter soliton is a solution of the KdV equation (5.20) and is described by the following parameters: characteristic width (diameter)

$$2a = \frac{4}{\sqrt{3}} H_0 h_0^{-1/2} \tag{5.25}$$

and propagation velocity

$$V_0 = c_0(1 + h_0/2) \ . \tag{5.26}$$

It is important to note for the subsequent discussion that the larger the amplitude of a KdV soliton, the greater its velocity and the smaller its width.

This soliton was discovered by John Scott Russell, a British naval engineer, in 1834, half a century before it was studied theoretically by Korteweg and de Vries. The story of Russell's discovery has been included in many books and reviews on solitons – see, for example, [5.26]. We reproduce it here for the reader's convenience.

One day Russell was watching a barge on the waterway between Glasgow and Edinburgh, pulled by a pair of horses. When the barge stopped abruptly, a small bell-shaped elevation of water separated from it and ran forward. It was about two feet high, and its width was approximately equal to the width of the waterway (30 ft). Fascinated by the sight, Russell followed the wave on horseback until it ceased to be discernible and disappeared in the meanders of the canal. This was the first consciously observed soliton in the history of science. Later Russell carried out numerous experiments with solitons in shallow water and established many of their properties, including the relations (5.25) and (5.26) and the ability to pass through each other in collisions. The experiments also revealed that the solitons under study were always elevations and never rarefactions.

Now let us return to the equation for rotating shallow water (5.4). This equation includes a scalar nonlinearity (the third term), similar to the KdV nonlinearity. If this nonlinearity prevails, it would be natural to expect, by analogy to the above KdV solution, that a soliton (if any) can only be an elevation on the surface since only in this case can the scalar nonlinearity counterbalance the dispersion. In the Rossby regime, however, an elevation is a geostrophically balanced anticyclone, hence only an anticyclone can be a Rossby soliton. In the case of a cyclone, which is a depression on the liquid surface, the nonlinearity and the dispersion are of the same sign and cannot balance out. Thus a Rossby cyclone cannot be a soliton in the regime given by (5.21). We shall now discuss this crucial fact in some more detail [5.43].

The buildup of a one-dimensional Russell-KdV soliton is due to a combination of two factors:

(a) positive quadratic nonlinearity (the term $(v\nabla)v$ in the Euler equation), meaning that the phase velocity ω/k of the wave increases with wave amplitude, and

(b) negative dispersion, meaning that ω/k decreases with increasing wavenumber k.

Both these factors are also present in (5.2). One might therefore expect that nonlinear Rossby waves can also be solitary waves – that is, solitons – rather similar to the solitary KdV wave. The fact that (5.2) describes a two-dimensional rotating shallow water rather than a one-dimensional case without rotation is of little consequence here.

The dispersion of gravity waves in nonrotating fluids is given by the following relation [5.44, 45]:

$$\omega^2 = gk\tanh(kH_0) , \tag{5.27}$$

the phase velocity of such waves being

$$\omega/k = (gH_0)^{1/2}\left(\frac{\tanh kH_0}{kH_0}\right)^{1/2} \tag{5.28}$$

The dispersion curves for these waves and for Rossby waves are shown in Fig. 5.4. The waves of both kinds display the following noteworthy common features:

(a) as the wavenumber tends to zero, the wave velocity tends to its maximum; for Rossby waves, the maximum value is V_R given by (5.9, 11, 12) while for gravity waves it is much larger: $c_0 = (gH_0)^{1/2}$;

(b) the dispersion is strong in the vicinity of the "characteristic" wavenumbers $k_x \approx 1/r_R$ and $k \approx 1/H_0$, the quantities r_R and H_0 playing the roles of mutually analogous dispersion parameters;

(c) the dispersion is negative in this region, making the phase velocity of the waves fall with increasing wavenumber.

It should be added that both kinds of waves are characterized by a "positive" nonlinearity, by which we mean that their velocities (for instance, V_R or c_0) increase with the amplitude of surface elevation.

The fact that the KdV soliton on a free fluid surface can exist only as a solitary wave of elevation is usually interpreted as follows. Consider a bell-shaped elevation on the surface of the liquid and examine whether it can, in principle, be a soliton. Such a perturbation on the surface can be constructed with a set of waves (a wave packet) in which the longer waves contribute mainly to the pedestal while the shorter ones are responsible for the steep slopes and the cap of the elevation. Because of the negative dispersion, the steep slopes associated with the shorter waves must propagate slower than the pedestal; on the other hand, being elevated, they must travel faster due to the nonlinearity. This gives rise to the opportunity for the dispersion spreading of the components to be counterbalanced by the nonlinearity, resulting in a solitary wave, or a soliton.

Now consider a similar perturbation on the surface, this time having the form of a depression. In this case, the steep slopes will travel slower than the pedestal due to both the dispersion and the nonlinearity. The nonlinearity not only fails to prevent the dispersion spreading of the packet but actually assists it. Consequently, a trough-like soliton is not feasible, in full agreement with Russell's experiments and observations.

The exact analytical solution for the problem of solitary wave in (nonrotating) shallow water, as mentioned already, was obtained by Korteweg and de Vries in 1895 (the history of the KdV equation and the preceding soliton-type solutions obtained by J. Boussinesq and Lord Rayleigh are discussed, for example, in [5.38, 39, 44, 46, 47]).

According to (5.26), the soliton velocity is somewhat higher than the maximum velocity of linear waves $(gH_0)^{1/2}$. This is a prerequisite of the soliton's stability: since its velocity is higher than that of any wave satisfying the relevant dispersion relation (5.27, 28), see Fig. 5.4b, it does not waste its energy on Cherenkov emission of such waves and hence does not decay. It exists in such a range of parameters where the wave dispersion is counterbalanced by the nonlinearity. That happens at "middle" wavelengths, $\lambda \gtrsim 2H_0$; the longer wavelengths are dominated by the nonlinearity and the shorter by the dispersion.

Now let us return to Rossby waves in the regime given by (5.21). The above-mentioned fact that they are similar to gravity waves by the nature of their dispersion and their scalar nonlinearity means that we may apply to the Rossby wave case all of the qualitative arguments which led us to the conclusion that a Russell (KdV) soliton in the case of nonrotating shallow water can exist only as an elevation. Since an elevation corresponds to an anticyclone and a depression to a cyclone for Rossby waves, we conclude that a Rossby soliton, if any, must be an anticyclone and not a cyclone. Similarly, we can come to the conclusion that the drift velocity of a Rossby soliton in the direction of Rossby wave propagation (that is, westwards) must be somewhat greater than the maximum linear wave velocity V_R, and that the characteristic radius of a Rossby soliton must correspond to intermediate wavelengths so that the effects of dispersion and nonlinearity compensate each other. This requirement, which is the counterpart of $\lambda > 2H_0$ for nonrotating fluid, implies $a > r_R$ coinciding with (5.21). All of the above qualitative considerations are confirmed in the rigorous theory of monopolar Rossby solitons [5.11, 12, 14–20].

Similar solitons must also exist in the Rossby regime of magnetized plasma. They are the drift solitons, solitary waves of plasma potential traveling in the same direction as the linear drift waves at a speed a little above the drift velocity V^*.

A scalar Rossby soliton (as well as a drift soliton in the magnetized plasma) is more or less round and its characteristic radius is somewhat greater than r_R (or r_L). According to the theory in the above references, the shape of a monopolar scalar (necessarily anticyclonic) Rossby soliton is universal: its elevation profile is described by

$$h = h_0 \cosh^{-4/3}(3r/4a) \tag{5.29}$$

and its characteristic radius is uniquely determined by its amplitude:

$$a \approx \sqrt{3} r_R h_0^{-1/2} \tag{5.30}$$

Figure 5.6 shows the profile obtained numerically in [5.12] which is close to that described by (5.29, 30). The soliton propagates "westwards" – counter to system rotation – at a velocity slightly greater than the Rossby velocity V_R.

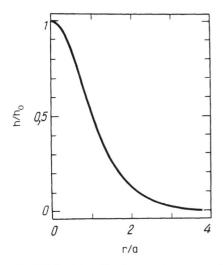

Fig. 5.6. Spatial profile of a smooth monopolar Rossby soliton: liquid surface elevation as a function of the distance from vortex center. From [5.12]

A monopolar soliton with the above properties was supposed to be a "purely wave" formation, carrying no particles of the liquid. According to the numerical studies, such solitons should collide elastically, that is, they should pass through each other like the classical Russell-KdV solitons [5.48, 49].

However, a "purely wave" Rossby soliton cannot be stationary: as we have mentioned at the end of Sect. 5.4, it must decay due to the inhomogeneity of the Rossby velocity over its meridional cross-section. This has been first shown in [5.21].

For a certain period after the appearance of [5.21] it seemed that there can be no monopolar Rossby soliton (in a rotating liquid with no flows). However, our experiments described in Chapt. 9 have shown that the Rossby soliton does exist. The condition for its existence was found to be a new qualitative feature, not taken into account in the theoretical studies cited above, namely the presence of a particle capture region in the central part of the soliton where the streamlines are closed. The Rossby soliton is then a "genuine" vortex. Such a structure can be implemented if the characteristic rotation velocity of the captured liquid in the soliton (the maximum value across its profile) exceeds the velocity of soliton drift as a body: $v_0 > V_{dr} \approx V_R$ – this is nothing else but (2.15).

We are now going to show that the necessity of the particle capture condition (2.15) for the existence of the Rossby soliton follows indeed from simple theoretical arguments (which was not noticed by the authors of the publications cited above).

Let us return to the linearization of (5.4) carried out earlier. Disregarding the last term in that equation, we had (in dimensionless units) $\omega(1+k^2)h+k_x V_R h = 0$ where k_x is the wavenumber along the x axis. Now take into account the last term in (5.4), demanding that after linearization it should not violate the Rossby wave dispersion (5.8). This means that the linearized value of the term, $2V_R k_x hy/R$, must be small compared to the smallest term of the last equation above, $\omega k^2 h$. Utilizing the relation $\omega \approx V_R k_x$, valid for the long waves in question ($k^2 r_R^2 \ll 1$), we obtain $2V_R k_x hy/R \ll V_R k_x k^2 h$. Hence, assuming $2y \approx a$ and $k \approx 1/a$, we come to the condition which in the dimensioned units has the form $a^3 \ll r_R^2 R$, or

$$a < (r_R^2 R)^{1/3} \tag{5.31}$$

The quantity

$$r_{IG} = (r_R^2 R)^{1/3} = (R/r_R)^{1/3} r_R \tag{5.32}$$

is known as the intermediate geostrophic radius. It follows from (5.31), in accordance with the conclusions of the theoretical studies [5.11, 14–17], that the size of our soliton should not exceed the intermediate geostrophic radius. Combining the condition (5.31) with (5.30) and with the equation of geostrophic equilibrium (2.11), we obtain $v_0 > 3V_R$ which is practically equivalent to the capture condition (2.15).

Thus a Rossby soliton cannot be a "pure wave"; it must necessarily be a "genuine vortex" carrying trapped particles. This effect (first discovered experimentally, as mentioned above) has been laid in the foundation of a new theory developed by G.G. Sutyrin [5.50–54] and based to a large extent on numerical calculation. As shown in that theory, a vortical soliton, contrary to a "pure wave" one, has no universal shape, its profile and size generally being independent of its amplitude: for a given amplitude, there can be many bell-shaped elevation solitons of different profile widths. A particular soliton is produced depending on the generation conditions and retains a "memory" of the area where the initial vorticity satisfied the capture condition (2.15). This "memory" is due to the invariance of the closed isolines of potential vorticity and follows from the Larichev theorem mentioned in Sects. 9.7, 10.2. These solitons have a quite definite shape only outside the particle capture area (where the shape is determined by (5.29, 30)); inside this area it is, generally speaking, arbitrary enough. The inner and outer soliton profiles are conjugated at the separatrix which separates the closed and open streamlines. Some higher derivatives of the vorticity are discontinuous here;[2] however, judging from the results of numerical calculations, this does not significantly affect soliton lifetimes – they are much longer (about two orders of magnitude) than those of linear Rossby wave packets [5.53–54].

One more essential fact should be noted: among the numerous possible vortical solitons of a given amplitude, only one is "smooth" (that is, all the derivatives of its vorticity are continuous everywhere), its profile being very close to that described by (5.29, 30). Clearly, rather special conditions are required for the realization of such a "rare" soliton.

An important distinction of the solitons "with memory" from the "pure wave" solitons is the different character of their collisions: the vortical Rossby solitons collide inelastically – they merge.

Thus the two theories of the monopolar Rossby soliton – "pure wave" and vortical – are so different that there should be no serious difficulty in choosing the correct theory by comparing them with the experimental data.

A vortical monopolar Rossby soliton is illustrated in Fig. 5.7, showing streamlines in the vortex frame which moves to the right with respect to the liquid. One of the streamlines, the separatrix, contains the point O where the velocity of liquid particles is zero. The area enclosed by the separatrix is the particle capture region.

Soliton geometry corresponds to the condition given by (5.21), and the scalar nonlinearity (that is, its balance with the dispersion) is mainly responsible for the formation of the soliton. In contrast to the "pure wave" soliton obtained under the assumption $h_0 \ll 1$, the conditions for the existence of a vortical soliton are restricted by a softer requirement $h_0 \leq 1$. In particular, a "smooth" soliton having the maximum amplitude ($h_0 \approx 1$) is feasible. It is interesting to note that the latter, too, has a profile close to that described by (5.29, 30), practically coinciding with what is shown in Fig. 5.6. The presence of captured particles is what distinguishes the Rossby soliton from the Russell-KdV soliton which is known to carry no particles of liquid along with it.

In connection with the condition (2.15), we should explain the ambivalent meaning of the term "vortex", as used here. Understood as a velocity curl carrier,

---

[2] The Larichev-Reznik dipolar soliton has a similar discontinuity of higher derivatives at its separatrix [5.6]. Since there are two "genuine" vortices, a cyclone and an anticyclone, inside the separatrix, the dipolar vortex is practically insensible to the last term of (5.4).

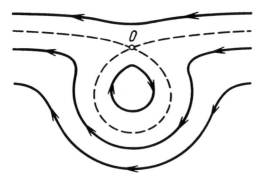

Fig. 5.7. Schematic of a monopolar Rossby soliton: streamlines in the frame which is drifting with the soliton along the parallel; *dashed curve* shows the separatrix enclosing trapped particles; the velocity of liquid is zero at point O

a vortex may be a formation of any amplitude, including a linear Rossby wave, since the particle trajectories in it exhibit vorticity under the influence of the Coriolis force. However, a "genuine" vortex must carry along captured particles of the liquid and should therefore have a large enough amplitude to satisfy the capture condition (2.15).

The above theory of the scalar monopolar Rossby soliton is compared with experimental data in Chaps. 9, 10. In particular, this theory is applied there to the interpretation of natural vortices in atmospheres and oceans.

The scalar soliton drifts in the same direction as linear Rossby waves (westwards) at a velocity greater than the Rossby velocity V_R. Therefore, like the vector soliton, it does not waste its energy on Cherenkov emission of linear Rossby waves propagating in the same direction. The absence of Cherenkov emission is a prerequisite for the soliton to be stationary. However, the following question might arise: could there be a direction of linear Rossby wave propagation such that the corresponding projection of soliton velocity coincides with the phase velocity of Rossby waves? Should such a direction exist, this would imply the possibility of Rossby wave radiation in that direction. It is, however, easy to see that there is no such direction. Indeed, the Rossby velocity along a direction at an angle θ to the parallel is the Rossby velocity along the parallel times $\cos\theta$. But the projection of the soliton velocity on that direction is its longitudinal velocity multiplied also by $\cos\theta$. This means that the soliton travels faster than any linear wave in any direction, that is, no Cherenkov emission of Rossby waves is possible. (This point was brought to our attention by Yu.A. Stepanyants [5.55].)

The drift velocity of the scalar Rossby soliton increases with its amplitude according to the following law [5.50–52]:

$$V_{dr} \approx V_R(1 + \kappa h_0) \tag{5.33}$$

where $\kappa > 0$. The physics behind (5.33) is quite clear: the soliton propagation velocity, like the Rossby velocity, increases with growing thickness of the liquid layer H where $H = H_0 + \delta H$, and $\delta H > 0$ for anticyclones. The law (5.33) is quite similar to the law (5.26) governing the Russell-KdV solitons.

For the smooth soliton of maximum amplitude, $\kappa \approx 0.6$; this soliton, like any other "smooth" one, has a low probability of realization. For the much more usual soliton "with memory", the value of κ depends on vortex profile [5.51, 52]; for the typical profiles observed experimentally, $\kappa \approx 0.2$ (Chap. 9).

The following reminder is appropriate here. We have deduced the necessity for the Rossby soliton to be a "genuine" vortex from the condition (5.31) demanding that the dispersion of Rossby waves be conserved. Thus the Rossby soliton features an organic combination of vortical and dispersive (wave) properties – it is a dualistic object (this is discussed also in Sect. 9.6).

Concluding this section, we would like to make an important remark. Sometimes [5.22–25] attention is drawn to the fact that, due to the above-mentioned "twisting" (caused by the meridional inhomogeneity of the Rossby velocity), a monopolar Rossby soliton must, in principle, be non-stationary even if the "vortex" condition (2.15) is satisfied. This is due to the fact that at sufficiently distant points of the soliton "tail" (the "wave" area) particle rotation is not rapid enough to eliminate the "twisting". In this connection, we note the following. The "twisting" will be substantially slowed down (that is, it will occur for a period longer than R/V_R) even if particle rotation velocity is as low as $\pi V_R a/R$. Since this value is much less than that corresponding to the separatrix of the vortex, only the far tail of the soliton will be subject to "twisting". It is clear that any substantial deformation of the vortex as a body can occur only in times much greater than R/V_R (which is confirmed by the numerical calculation [5.14–17]). It is then natural to ask whether one ought to take into account such a weak source of non-stationarity? We believe that one ought not, since, as is known, factors of weak non-stationarity can also be pointed out for the Russell-KdV [5.26], Larichev-Reznik [5.6], Hasegawa-Mima [5.7], and other solitons.

In the natural conditions, a free soliton must in principle be non-stationary even due to the viscosity of the medium, and against that background the above factor of weak non-stationarity may be unnoticeable. In planetary atmospheres, viscosity is counterbalanced due to vortex generation by zonal flows. The flows will also compensate for the above-mentioned weak factor of the "wave" non-stationarity of the vortices.

5.6 Viscous Damping of Rossby Vortices

The structures discussed above are subject to viscous damping. The characteristic time of this process can be assessed with the following simple arguments. It is known that the velocity of a free (unpumped) plane-parallel flow in a layer of rotating incompressible liquid of thickness H_0 is damped due to the friction against the bottom with a characteristic (Eckman) time τ_E given by (4.8) [5.4]. Rossby vortices in free-surface shallow water have a characteristic decay time

τ_ν greater than τ_E because, as contrasted to the case of a plane-parallel flow under a "solid lid" (for which (4.8) is written), they possess not only the kinetic energy W_k but also the so-called available potential energy W_p. Their decay time is [5.4]

$$\tau_\nu = \tau_E(1 + W_p/W_k) \ . \tag{5.34}$$

Taking into account that $W_k \approx H_0 v_0^2/2$ and $W_p \approx g^*(\delta H)_0^2/2$ (the liquid density has been assumed to be unity) and expressing $(\delta H)_0^2$ in terms of v_0^2 by means of (2.11), we have, independently of the vorticity sign (cyclone or anticyclone), that is, independently of what is the deviation of the free surface (elevation or depression):

$$\tau_\nu \approx \tau_E(1 + Ka^2/r_R^2) = \tau_E(1 + KF) \ . \tag{5.35}$$

The factor K entering (5.35) is of the order of unity and needs some refinement. The numerical calculations carried out by A.V. Khutoretsky [5.37] show that $K \approx 1/2$ for a round vortex of Gaussian profile; a is understood here as the distance from vortex axis to the point where the linear rotation velocity is the highest. We shall use this value of K later.

Equation (5.35) has an interesting and somewhat unexpected implication. The time τ_ν consists of two components. The first is related to vortex kinetic energy and is proportional to H_0. The second is related to the available potential energy; it depends on vortex amplitude and *does not depend* on H_0. Therefore, as $H_0 \to 0$ (but, of course, $H_0 > l_E$), the value of τ_ν tends to a *finite* limit. Indeed, as $H_0 \to 0$, the second term in the parentheses in (5.35) becomes very large and one may neglect the unity; the quantity H_0 then drops out of the expression for τ_ν. We shall take this effect into account, in particular, when comparing the theory with the experimental data.

Comparing the times τ_ν for cyclones and anticyclones, the following important circumstance must be considered. Strictly speaking, the fact that the Rossby number is finite should be taken into account, that is, f in (4.8) should be replaced with $f + \omega$ for cyclones and with $f - \omega$ for anticyclones, where ω is the characteristic frequency of vortex proper rotation. In addition, the layer thickness H_0 should be replaced with $H_0(1 - h_0)$ for cyclones and with $H_0(1 + h_0)$ for anticyclones. However, these corrections to H_0 and to f largely cancel one another, so that the values of τ_ν given by (5.35) for cyclones differ from those for anticyclones by no more than 30%.

Additional explanations for some of the material discussed in this chapter are given in Supplements 5.1–4.

6. Experimental Configurations

In this chapter we describe our experimental configurations which have been used to produce for the first time Rossby solitons simulating both the largest vortices in atmospheres and oceans and drift solitons in plasma, and to generate nonlinear spiral waves simulating the spiral structures of galaxies. We begin by outlining the requirements an experimental configuration should comply with to ensure that the solitonic features of Rossby vortices, if any, could be observed so that Rossby solitons could be identified positively.

6.1 Geometry and Rotation Regime of the Vessels for the Observation of Rossby Solitons

The free surface of a liquid rotating in a gravitational field at a constant angular velocity Ω around the vertical axis (Fig. 6.1) assumes a parabolic shape described by

$$\eta = \Omega^2 \varrho^2 / 2g , \tag{6.1}$$

which follows directly from the equilibrium condition $\Omega^2 \varrho \cos \alpha = g \sin \alpha$ for the forces tangent to liquid surface, whence

$$\tan \alpha = d\eta / d\varrho = \Omega^2 \varrho / g \tag{6.2}$$

where η and ϱ are respectively the vertical and radial cylindrical coordinates of a surface point, their origin fixed at the lowest point of the surface.

A layer of rotating free-surface liquid can serve as a model of a homogeneous ocean or a planetary atmosphere if its thickness is constant and small enough for the shallow water approximation to be valid. This implies that the bottom of the vessel, rotating together with the liquid, should be approximately parabolic in shape, close to that described by (6.1). To be more precise, if the bottom and the surface have identical shapes, the liquid layer will be uniformly thick if measured along the vertical axis η, not along the normal to its surface ($H_\eta = H_0 / \cos \alpha =$ const). Thus to obtain a uniform layer thickness H_0 along the local normal, the vessel rotating at a rate Ω_0 should be somewhat less sloped than as implied by (6.1) for $\Omega = \Omega_0$. Such vessels are used in all of our experiments [6.1–12] (except for a single control experiment, see below); for brevity, we shall refer to

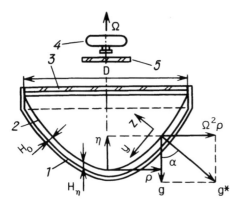

Fig. 6.1. Geometry of rotating shallow water and the experimental configuration for studying Rossby vortices in their free-travel regime: (**1**) vessel with nearly parabolic bottom, "paraboloid"; (**2**) surface of the liquid which spreads evenly over the bottom of the rotating vessel in an equilibrium layer of thickness H_0; (**3**) transparent lid shutting off "wind" (preventing the rotating liquid from friction against the air); (**4**) camera; (**5**) red filter. The vessel and the camera rotate at angular rate Ω; D is vessel diameter

them as paraboloids (Fig. 6.1). The bottom shape of each of these paraboloids is uniquely determined by two parameters: the nominal value of H_0 for which it is intended and the rate Ω_0 at which it is to be rotated in order to obtain, with a certain "nominal" total amount of liquid, an equilibrium layer having the same thickness H_0 over the entire vessel.

In a rotating paraboloid (Fig. 6.1), the liquid layer is under the joint action of two forces, the gravity force and the centrifugal force due to overall system rotation; in equilibrium, the net acceleration is directed perpendicular to the liquid surface and has the magnitude

$$g^* = g/\cos\alpha . \tag{6.3}$$

Then, with $H_0 = $ const, the Rossby-Obukhov radius in the paraboloid is

$$r_R = (gH_0)^{1/2}(2\Omega_0\cos^{3/2}\alpha)^{-1} , \tag{6.4}$$

and the Rossby velocity, according to (5.11), is

$$V_R = \frac{g^*}{R}\left|\frac{\partial(H_0/f)}{\partial\alpha}\right| = (1/2)H_0\Omega_0\sin\alpha \tag{6.5}$$

where R is the meridional curvature radius of the liquid surface:

$$R = g\Omega_0^{-2}\cos^{-3}\alpha . \tag{6.6}$$

One more point, important for the experiment, should be mentioned here. If the total amount of liquid remains nominal but the paraboloid is rotated at an angular rate Ω exceeding the value Ω_0 at which $H_0 = $ const by a quantity

$\Delta\Omega$ then the layer thickness will exhibit a meridional gradient directed outwards (or inwards, if $\Delta\Omega$ has the opposite sign). According to (5.11), the Rossby velocity V_R will then have a different value since now $H_0 \neq$ const. The simplest expression for V_R is obtained for that parallel of the paraboloid which lies at the distance $\varrho = D/2\sqrt{2}$ from the rotation axis, where D is the diameter of the vertical cylindrical envelope of the paraboloid. This was chosen as the working parallel in our experiments on vortex generation. Layer thickness does not change here with changing Ω if $\Delta\Omega$ is small enough, and

$$V_R = \frac{H_0 \Omega_0 \sin \alpha}{2} \left[1 + \frac{\Delta\Omega}{\Omega_0} \frac{D/H_0}{1 + \Omega_0^4 D^2/8g^2} \right] . \tag{6.7}$$

We see that the value of V_R increases if $\Delta\Omega > 0$ (the vortex drifts "westwards" relative to the paraboloid faster than when $\Delta\Omega = 0$); if $\Delta\Omega < 0$, the Rossby velocity decreases and may even change its sign (the vortex will then drift "eastwards").

The quantity of liquid in the vessel may be changed from experiment to experiment; the layer thickness H_0 in the vicinity of the working parallel can then be maintained the same, if necessary, by a corresponding change in the paraboloid's angular rotation rate.

Should the vessel be made an *ideal* paraboloid, the Rossby velocity would be zero for the angular rotation rate Ω_0 at which the shape of the liquid surface is identical to that of the bottom (that is, $H_\eta = $ const), since, in accordance with (5.9, 11),

$$V_R = g^* \frac{\partial}{\partial y} \left(\frac{H_0/\cos \alpha}{2\Omega_0} \right) = \frac{g}{f} \frac{\partial H_\eta}{\partial y} . \tag{6.8}$$

One of the distinguishing characteristics of Rossby vortices, their drift along the parallel, would vanish in that case.

If, however, the condition $H_\eta = $ const is violated by shifting the angular velocity of the ideal paraboloid by $\Delta\Omega$, then with the aid of (5.9, 11) the Rossby velocity can be shown to be

$$V_R = \varrho\Delta\Omega \tag{6.9}$$

where ϱ is the radius of the parallel along which the vortex is drifting. The vortex drifts in the direction of vessel rotation if $\Delta\Omega < 0$ and counter to it if $\Delta\Omega > 0$.

To complete our picture, let us also discuss the following point. Evidently, other factors being equal (with given values of Ω and H_0), one can generally influence the value of V_R by changing the vessel meridional curvature at the desired radius ϱ. Since in such a case the free surface profile and hence also the free surface curvature (6.6) and the meridional gradient of the Coriolis parameter will all remain the same, the Rossby velocity change will be due only to the change in the meridional gradient of H_0 – according to (5.11). It is easy to imagine a situation where the vessel bottom is so "unparallel" to the free surface that the term containing $\partial H_0/\partial y$ in the expression (5.11) for V_R becomes much larger than that containing $\partial f/\partial y$. Under such conditions, the meridional inhomogeneity of the Coriolis parameter contributes only negligibly to the overall β-effect which may nevertheless be quite significant. Such a geometry is typical for many experimental studies. For example, a traditional and very fruitful method of creating the β-effect in experiments with rotating liquids is to use a vessel whose bottom has the shape of a cone with its vertex directed upwards. This so-called "inclined" bottom, even with no free surface (when the liquid is covered with a flat lid), creates a strong radially directed gradient of liquid depth, making it possible to achieve a pronounced β-effect of the same sign as that existing on planets. An example of such experiments will be discussed in Sects. 7.1, 5 (see also Supplements S5.2, 3). Here we would just like to mention the experiments described in [6.13] where the same principle of creating the β-effect with a strong gradient of H_0 has been used for studying wave phenomena in rotating free-surface liquid. The vessel bottom was exactly parabolic in these experiments but quite different from the shape of the free liquid surface due to an appropriate choice of the angular rate. In this connection, we must explain the following important circumstance that should be kept in mind when setting up laboratory experiments for studying Rossby solitons.

For a *scalar* Rossby soliton to exist in a real experimental configuration, its diameter (projected onto the horizontal plane) must be much smaller than the radius of the parallel at which it is located: $2a\cos\alpha \ll \varrho$. Since for this type of soliton, according to (5.30), certainly $a \gtrsim 2r_R$, the above necessary condition implies $4r_R\cos\alpha \ll \varrho$, whence, utilizing (6.1–4), we obtain $H_0/\eta \ll (\cos\alpha)/2$. To satisfy the last inequality, the vessel bottom must have a shape sufficiently close to the shape of the free liquid surface. This means that the geometry of the above-mentioned experiments where the β-effect is induced by a strong gradient of liquid layer thickness (in particular, when a vessel with an inclined bottom is used), rather than by a gradient of the Coriolis parameter as it occurs on planets, is unsuitable for creating scalar Rossby solitons.

Thus a parabolic shape of vessel bottom appears as an obvious choice. (The possibility of observing Rossby solitons in parabolic vessels has been mentioned in [6.14], in the following words: "These vortices can be easily produced in the laboratory in a shallow liquid in a rotating container with a parabolic shape of the bottom".) However, the problem of selecting the appropriate geometrical parameters and rotation velocity of the paraboloid calls for a special attention. To make the identification of Rossby solitons possible, the paraboloid must comply with certain rather stringent requirements. Indeed, to prove that a Rossby vortex is a soliton, one has to show that it is immune to dispersion spreading inherent in a linear wave packet, or at least that the time of its dispersion spreading is much longer than that of a linear wave packet. Thus one needs a vessel where the dispersion properties of Rossby vortices could be manifested clearly enough. This results in the requirement that the vessel bottom be sufficiently steep at its working parallel. Otherwise, the dispersion properties of the nonlinear structures in question are so weak that they are virtually indiscernible during vortex lifetime limited by such an inevitable factor as structure damping due to the viscosity of the working liquid. If the paraboloid were not steep enough, the lifetimes of the structures would not depend on whether they suffer dispersion decay (like linear wave packets) or not (like solitons). The possibility to identify the structures as Rossby solitons would then be precluded in principle.

A quantitative estimate for the required steepness of the paraboloid can be obtained by demanding that the characteristic dispersion spreading time τ_1 of a linear Rossby wave packet be as small as possible. It is still admissible, for example, if the value of τ_1 is comparable to the characteristic time τ_ν of the inevitable viscous damping, but $\tau_1 \gg \tau_\nu$ is out of question. For a paraboloid with $H_0 = \text{const}$, according to (5.16, 6.2, 4, 5), we have the following expression for the minimum dispersion spreading time of a linear Rossby wave packet, corresponding to $a \approx r_R$:

$$(\tau_1)_{\min} = 8\varrho(g H_0)^{-1/2}(1 + \tan^2 \alpha)^{5/4} \tan^{-2} \alpha \ . \tag{6.10}$$

The smallest value of $(\tau_1)_{\min}$ is obtained when

$$\tan \alpha = 2 \ . \tag{6.11}$$

Since α is the inclination angle of the tangent to the vertical cross-section of the paraboloid with respect to the horizontal, condition (6.11) means that the paraboloid must be sufficiently steep. It is seen from (6.10) that if $\tan \alpha > 1$ then the time $(\tau_1)_{\min}$ depends weakly on $\tan \alpha$, so there is no need to satisfy (6.11) *rigorously*. (When $\tan \alpha$ is too large, the necessary condition $D/2 \gg a > r_R$ becomes difficult to satisfy since r_R grows rather quickly with α by virtue of (6.4).) But if $\tan^2 \alpha$ is much less than 1, then $(\tau_1)_{\min}$ changes almost as $\tan^{-2} \alpha \propto \Omega_0^{-4}$. This implies that a paraboloid that is not steep enough at its working parallel is totally unsuitable for the experiments discussed here. We use two paraboloids which well satisfy the above criteria. One of the vessels (the smaller one) has diameter $D = 28$ cm, $\Omega_0 = 10.85$ rad/s, and $\tan \alpha = 1.20$ in the working area. The other (larger) vessel has $D = 73$ cm, $\Omega_0 = 7.48$ rad/s, and $\tan \alpha = 1.43$. The heights of the vessels are approximately equal to their respective outer radii $D/2$. With these parameters, the value of $(\tau_1)_{\min}$ as given by (6.10) is within 10% of the minimum corresponding to the condition (6.11). It was found, however, that even with this, nearly optimum, experiment geometry the time $(\tau_1)_{\min}$ is practically very close to the characteristic viscous damping time τ_ν if water is used as the working liquid. (The use of liquids other than water either does not improve the situation or requires a much more complicated experiment. Moreover, even if the viscosity of the working liquid were made negligibly small, another significant cause often limiting vortex lifetimes would remain – the vortices drift along the meridian and eventually collide with vessel walls). This means that if the paraboloids were not steep enough – so that $\tan^2 \alpha$ would be several times lower at the working parallel – any attempts of identifying

Rossby solitons in the experiment, not to mention attempts of studying their motion, would be doomed to fail.

The parameters of the paraboloids used in our experiments are given in Table 6.1. Additional details on the experiments can be found in [6.1–12].

Table 6.1. Parameters of the paraboloids

	Small paraboloid	Large paraboloid
Ω_0, rad/s	10.85	7.48
$\eta(\varrho)$, cm	$6.00 \cdot 10^{-2}\varrho^2$	$2.85 \cdot 10^{-2}\varrho^2$
D, cm	28	73
Radius of working parallel, cm	10	25
$(H_0)_{min}$, cm	0.3	0.4
$(H_0)_{max}$, cm	1.2	5.5
At the working parallel, for $\Omega = \Omega_0$ and $H_0 = $ const:		
H_0, cm	0.5	1.0
R_v, cm	32	93
r_R, cm	2.0	4.8
V_R, cm/s	2.1	3.1
$(\tau_1)_{min}$, s	7.6	12.6

Notations:
Ω_0 - rotation rate corresponding to equilibrium liquid layer of uniform thickness H_0; $\eta(\varrho)$ - equation of the free surface of liquid rotating at a rate $\Omega = \Omega_0$; D - maximum diameter of paraboloid; the radius of the working parallel is approximately $D/2\sqrt{2}$; R_v - meridional curvature radius of the vessel; $(\tau_1)_{min} = 8r_R/V_R$ - minimum dispersion spreading time of a linear Rossby wave packet

The optimization of the experimental configurations as described above made it possible to observe the dispersion behavior of the vortices in question. This behavior includes a pronounced cyclone-anticyclone asymmetry of long-lived Rossby vortices – a phenomenon similar to what is observed in the atmospheres of Jupiter and Saturn. Our experiments indicate that the underlying cause of this phenomenon is that the observed Rossby cyclones with $a > r_R$ undergo a rather quick dispersion decay in a characteristic time not larger than (6.10). On the other hand, the anticyclones with $a > r_R$ seem to show no dispersion spreading at all and appear to be Rossby solitons.

To conclude this section, an important methodological remark should be made. A theoretical study has shown that the influence of capillarity effects upon Rossby waves (or vortices) can be totally disregarded under the conditions of our experiments (as well as in any reasonably conceivable laboratory environment) [6.15]. The criterion for capillarity to have no influence on Rossby wave dispersion, obtained in [6.15], is $r_R^2 \gg (\lambda_0/2\pi)^2$. Here λ_0 is the so-called capillarity length, that is, the characteristic scale at which capillarity is manifested in the absence of system rotation (for water, $\lambda_0 = 1.7$ cm). Applying this criterion to the conditions of our experiments (see, in particular, Table 6.1), it is easy to see that it is satisfied for any conceivable experimental configuration (at least for liquids with no vertical stratification).

6.2 Techniques for Local Generation of Rossby Vortices

We shall now describe the experimental techniques we use for creating Rossby vortices in a layer of shallow water rotating as a single body (without counterflows). In all of these experiments, a local perturbation of shallow water is produced with some kind of pulsed "pumping generator". The perturbation then develops by itself into a vortical structure, existing subsequently in the "free-travel" regime.

The techniques are illustrated schematically in Fig. 6.2a–d. In the *first* of them (Fig. 6.2a), the perturbation is created by rotating a pumping disk placed flush with the bottom (similar to how the counterflows are produced in the experiments described in Sect. 4.1.2). The direction, duration, and angular rate of disk rotation are varied in the experiments. Disk diameter is usually close to the diameters of the Rossby vortices to be studied.

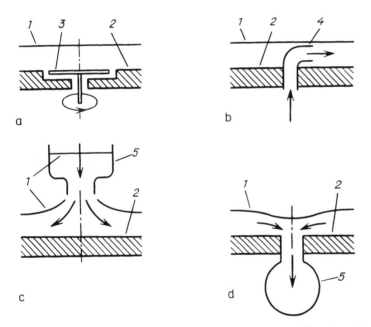

Fig. 6.2. Techniques "a", "b", "c" and "d" for local generation of Rossby vortices: (**1**) free surface of the shallow water; (**2**) surface of the bottom; (**3**) pumping disk; (**4**) tube; (**5**) auxiliary reservoir

In the *second* technique (Fig. 6.2b), a jet of liquid is injected into the layer of shallow water along the vessel bottom at the working parallel by means of a thin-walled tube, several times smaller in diameter than the layer thickness H_0. Jet direction, velocity, and the amount of injected liquid can be varied.

In the third technique (Fig. 6.2c), a controlled amount of liquid is poured down into the layer from above, along the direction of the local vertical, from a supplementary reservoir rotating with the vessel. The outlet of the reservoir is rather wide (about 3.5 cm in diameter) so that the liquid can be poured down in a quiescent ("laminary") manner in a short enough time. Subsequently, spreading symmetrically over the bottom, the liquid is twisted by the Coriolis force and forms an anticyclone with a self-consistent elevation of the free surface.

The same principle is used in the *fourth* technique (Fig. 6.2d), which provides a reliable and reproducible source of monopolar cyclones. A hole in the vessel bottom is opened at a predetermined time and some of the working liquid is drained into a supplementary reservoir. The depression in the shallow water thus created begins to fill with liquid coming from the adjacent areas, twisting on its way in the cyclonic direction. The parameters of the produced cyclone depend on how the liquid is drained, that is, on the volume of the supplementary reservoir and the diameter of the sinkhole.

Henceforth, in describing the experiments, we shall refer to the above ways of producing vortices as techniques "a", "b", "c", and "d", in accordance with Fig. 6.2. The choice of a technique depends on the specific goal of the experiment: first, on the vortex type to be studied (cyclonic or anticyclonic, monopolar or dipolar), and second, on the working thickness of the shallow water.

The experiments have shown, for instance, that techniques "a" and "b" generally lead, at the first stage, to the generation of coupled (dipolar) vortices. If the layer thickness H_0 is small enough, only anticyclones survive after some time in these dipoles, so with small H_0 techniques "a" and "b" can be used for studying the behavior of anticyclones. However, with relatively large H_0, when cyclone lifetimes are not so short, techniques "c" or "d" should be used for creating monopolar vortices.

Techniques "a" and "b" have been used in all our experiments with the small paraboloid and in some experiments with the large one. Techniques "c" and "d" have been used only with the large paraboloid.

6.3 Devices for Generating Rossby Vortices by Counterflows

To produce geostrophic counterflows in rotating shallow water, imitating zonal flows in planetary atmospheres and sheared flows in magnetized plasma, we use in our experiments the method described in part in Sect. 4.1.2 (Fig. 4.1) [6.4–12, 16]. The feature not mentioned in Chap. 4 is that the separation between the rotating rings which produce the counterflows can be varied from one experiment to another, thus providing control over the characteristic width δ of the transverse counterflow velocity profile. More precisely, in some experiments we set

$$\delta > r_R \ , \tag{6.12}$$

which corresponds to the conditions under which the largest atmospheric vortices on the giant planets, including the JGRS, are generated, while in other experiments

$$\delta < r_R , \tag{6.13}$$

which corresponds qualitatively to the generation of such vortices as, for example, Jupiter's Brown Ovals and those in the terrestrial atmosphere.

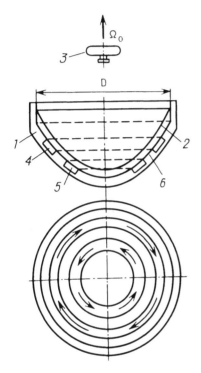

Fig. 6.3. Experimental configuration for producing Rossby vortices (solitons) by zonal flows with smooth velocity profiles: (**1**) vessel with nearly parabolic bottom; (**2**) surface of the liquid which spreads evenly over the bottom as the vessel rotates; (**3**) camera; (**4, 5**) counter-rotating rings in the bottom, creating zonal counterflows in the liquid (*arrows* in the plane view indicate flow directions with respect to the vessel for an anticyclonic velocity jump); (**6**) bottom zone between the moving rings (3 cm wide along the meridian). The vessel and the camera rotate at angular rate $\Omega_0 = 10.85$ rad/s; $D = 28$ cm

Three configurations based on the small paraboloid described in Sect. 6.1 (Table 6.1) are used in the experiments with zonal flows. The first of them differs from that shown in Fig. 4.1 only in that the sign of the counterflow curl can be changed from one experiment to another, allowing one to produce either cyclones or anticyclones. The second one (Fig. 6.3) is essentially the same, but with the rings separated by 3 cm (along the meridian). In the third one (Fig. 6.4), ring separation is wider (11 cm), and the rings can be rotated independently of each other; the inner "ring" in this geometry is the entire central area of the vessel, 10 cm in diameter. The most interesting regimes of vortex generation are implemented in the configuration shown in Fig. 6.4. In this configuration, where the condition (6.12) is satisfied by an ample margin, a stationary solitonic model for the generation of such natural vortices as the JGRS has been created successfully.

An important point concerning such simulations should be reiterated. In our experimental devices the shallow water layer rests on a solid bottom. Therefore one might ask if the experimental configurations are adequate for the actual conditions under which the atmospheric vortices of Jupiter and Saturn exist. As indicated in Sect. 2.2, the "bottom" of the Jovian atmosphere may be assumed to lie about 1000 km beneath the upper cloud layer: hydrogen, the main component of the Jovian atmosphere, is in the liquid state at that level. This depth is quite small compared to the horizontal dimensions of the JGRS vortex (12,000 km × x 25,000 km). So one may safely assume that not only the upper cloud layer of Jupiter (several dozen kilometers thick) but also the entire Jovian atmosphere is actually "shallow water" resting on a liquid bottom which is physically equivalent to the solid bottom of our experimental devices. Thus the shallow water experiments are quite adequate and proper for the simulation of such natural vortices as the JGRS.

As a matter of fact, even should one assume the thickness of Jovian atmosphere to be 25,000 km (the distance to the supposed solid core of metallic hydrogen), one would still remain within the limits of a two-dimensional structure, according to the well-known Taylor-Proudman theorem [6.17] which states that a rapidly rotating system tends to exhibit two-dimensional behavior.

6.4 Devices for the Simulation of Galactic Spiral Structures

The installations described in this section are, in principle, similar to those used to simulate the generation of atmospheric and plasma vortices by counterflows. The main component of the experimental configurations (Fig. 6.5) is a round vessel, the central part of its bottom (the "core", angular rate Ω_1) rotating faster than the outer part (the "periphery", angular rate Ω_2). The vessel is made in two different versions. The first of them (Fig. 6.5a) has core diameter $2\varrho_0 = 8$ cm and periphery diameter $D = 30$ cm [6.18]. The outer part of this vessel is either kept stationary or rotated at a relatively low rate, so that the Rossby-Obukhov radius r_R as defined by (6.4) is only four times smaller than the vessel radius. With such a low angular rotation rate the parabolic free surface of the liquid is almost flat, therefore the periphery of this version of the vessel is either flat or has the shape of an obtuse cone with its surface inclined to the horizontal at an angle of about 2°. The inner part of the vessel (the core) is also cone-shaped, with the cone generatrix inclined at 42.5° to the horizontal.

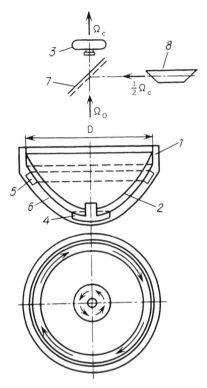

Fig. 6.4. Experimental configuration where a Rossby autosoliton was produced: (**1**) vessel with nearly parabolic bottom rotating at angular rate $\Omega_0 = 10.85$ rad/s; (**2**) surface of the liquid which spreads evenly over the bottom as the vessel rotates; (**3**) camera rotating at an adjustable angular rate Ω_c; (**4**) ring rotating faster than the vessel; (**5**) ring rotating slower than the vessel; (**6**) bottom zone between the moving rings (11 cm wide along the meridian); (**7**) semi-transparent mirror; (**8**) rotoscope based on a Dauvet prism rotating at angular rate $\Omega_c/2$; $D = 28$ cm

The second version of the experimental device (Fig. 6.5b) consists of two paraboloids and is twice as large as the first: core diameter $2\varrho_0 = 16$ cm, periphery diameter $D = 60$ cm [6.19]. Both parts of the bottom in this vessel are shaped in such a way that the working liquid, brought into a steady rotation by the bottom, is spread over the vessel in a thin layer of approximately constant thickness. A velocity jump is formed in the liquid layer between the core and the periphery. Its radius is ϱ_0 and its initial width is evidently close to the layer thickness H_0. We thus obtain a model of a galactic disk in the form of a shallow water layer whose velocity profile is similar to that of the galaxies which display a velocity jump between the core and the periphery (Fig. 2.13).

The experiments have been carried out in the following range of system parameters: 1.8 rad/s$\leq \Omega_1 \leq 42$ rad/s; $0\leq \Omega_2 \leq 3.6$ rad/s; 0.15 cm$\leq H_\eta \leq$ 0.4 cm. The respective values of the Mach number M$= (\Omega_1 - \Omega_2)\varrho_0(gH_\eta)^{-1/2}$

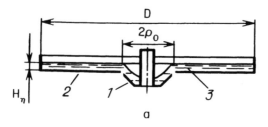

a

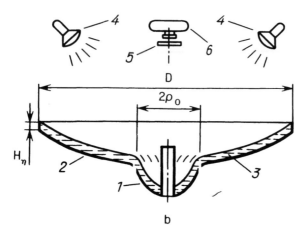

b

Fig. 6.5. Experimental configurations for simulating the mechanism which generates spiral structures in galactic gaseous disks: (**1**) core; (**2**) periphery; (**3**) layer of shallow water; (**4**) incandescent lamps; (**5**) red filter; (**6**) camera. Clockwise rotation in the plane view at angular rates Ω_1 (core), Ω_2 (periphery), and Ω_c (camera). (**a**) Configuration 1: $D = 30$ cm, $\varrho_0 = 4$ cm; (**b**) Configuration 2: $D = 60$ cm, $\varrho_0 = 8$ cm

and the ratios Ω_2/Ω_1 and $2r_R/D$ then ensure physical similarity to the galactic conditions: in the experiments, $0.5 \leq M \leq 12$; $\Omega_2/\Omega_1 \leq 0.2$; $2r_R/D \lesssim 0.1$. The kinematic viscosity of the working liquid is varied in the range 0.02 cm^{2}s$^{-1} \leq \nu \leq 0.5$ cm^{2}s^{-1} (the largest viscosity is 50 times that of water). The experiments carried out with both vessels yield results which are in good agreement with each other.

6.5 Diagnostic Techniques

Two diagnostic techniques are used for assessing the principal characteristics of the structures under investigation. One of them gives the surface density distribution, that is, the picture of the varying thickness of the liquid layer. The other shows the movements in the liquid and provides the flow pattern (the perturbed velocity field) in shallow water. Both techniques allow visual monitoring and photographic registration of the patterns.

When variations of shallow water thickness are to be studied, a vessel with a white bottom is used and the working liquid is water tinted with $NiSO_4$. The vessel containing this translucent greenish solution is illuminated from above with incandescent lamps. In visual observation or in black-and-white photographs taken through a red filter, the elevated areas of the free surface look darker than depressions against the white vessel bottom. This technique allows one to obtain thickness profiles of the structures under investigation in absolute units. In order to achieve this, the negatives are photometered and the whole illumination-to-film channel is calibrated by taking test pictures of the unperturbed liquid layer whose thickness is measured directly.

Motions in the liquid are studied by the "traditional" method – that is, by using test particles (white paper circles 1 or 2 mm in diameter) floating on the surface. The velocities of the test particles are determined from the lengths and directions of their tracks in the photographs (with a known exposure time). These particles are clearly discernible against the relatively dark background of tinted water (with a white bottom), making it possible to employ both techniques simultaneously. When the experiments are carried out with untinted water, a vessel with a black bottom is used (to make a better contrast with the test particles) and only the flow patterns can be observed.

The optical axis of the photographic camera is aligned with the vessel rotation axis. The camera can be rotated about the system axis at a controllable angular rotation rate Ω_c. This rate, depending on the goals of an experiment, may be either equal to the vessel rotation rate or chosen so that the structures under investigation, moving with respect to the vessel, would stand still relative to the camera. A rotoscope made on the basis of a Dauvet prism is used in order to facilitate the setting of the camera's angular rotation rate and the visual observation of the structures.

7. Laboratory Simulation of Rossby Vortices and Solitons in Planetary Atmospheres and Oceans

The experiments described in this chapter and in Chaps. 9–11 have been performed in the following succession. First, Rossby vortices and solitons were studied in a liquid rotating as a single body. In these experiments, the vortex under investigation was generated with a local pulsed (single-action) source and then propagated along the parallel in the free-travel regime through the rotating parabolic layer of shallow water. The lifetime of the vortex was limited by viscosity, apart from any other factors [7.1–3]. Next, experiments were carried out on generating Rossby vortices with stationary, axially symmetrical geostrophic counterflows. These experiments produced steadily drifting chains of vortices, from ten vortices in a chain when the flow velocity was low to one when it was high. This second series of experiments is of particular interest since it fits much better with the actual process of Rossby vortex generation by zonal flows in planetary atmospheres, therefore this series will be the first to be described. On the other hand, to elucidate the physical nature of the vortices in question one has to investigate them in their free-travel regime. Without such a study, it is impossible to tell whether there are any Rossby solitons among the observed vortices and whether they simulate plasma drift vortices and solitons. This series of experiments is described in Chaps. 9–11.

7.1 Generation of Rossby Vortex Chains by Zonal Counterflows in Rotating Shallow Water. The Cyclone-Anticyclone Asymmetry

It is well known that fluid flows with velocity shears are generally unstable. A typical example of such an instability, associated with the names of Kelvin and Helmholtz, has been studied in the experiments discussed in Sect. 4.1. It corresponds to a flow regime satisfying (6.13) which is close to a tangential discontinuity in the velocity profile. The separation between the rings which produce the counterflows is about 1 mm. The experiments indicate that this regime is very efficient in generating vortices – with either sign of the counterflows' velocity curl. With a cyclonic curl, cyclone chains are generated (Fig. 4.2), and with an anticyclonic curl, anticyclone chains are produced (Fig. 7.1). Thus in this regime – with an abrupt profile of flow velocity – vortex generation looks qualitatively the same for both polarities: a pattern resembling the well-known Kelvin's "cat's eyes" is created.

Fig. 7.1. Flow pattern in free-surface shallow water in the case of a sharp velocity profile and anticyclonic vorticity of counterflows (experimental configuration as in Fig. 4.1; the white circle passing through anticyclone centers is the line of tangential velocity jump)

An entirely different situation is observed with a smooth velocity profile satisfying (6.12). In this regime, flow instability will generate a chain of large-scale vortices (larger than r_R) only if the flow velocity curl is anticyclonic. The vortices produced are accordingly anticyclones (Fig. 7.2a) and look similar to those in Fig. 7.1. If the curl sign is switched to cyclonic, no large-scale vortices are generated (Fig. 7.2b). Thus a very pronounced cyclone-anticyclone asymmetry is observed when the counterflow profile is smooth. We attach a fundamental physical importance to this phenomenon because it obviously corresponds to the cyclone-anticyclone asymmetry in the large-scale, long-lived Rossby vortices in the atmospheres of the giant planets (Sect. 2.2). Therefore the investigation of the mechanism of this phenomenon will be one of the goals in the discussion to follow.

It is important to stress in this connection that there is no cyclone-anticyclone asymmetry when the flow profile is close to a tangential discontinuity (velocity profile width $\delta \to 0$) and the instability has the maximum growth rate given (for $c_s \to \infty$) by the well-known relation [7.4, 5]

$$\gamma \approx 2\pi u/\lambda \tag{7.1}$$

where $2u$ is the magnitude of the flow velocity jump and λ is the wavelength. When the flow profile has a finite width δ, the linear growth rate of the K-H instability in question, neglecting the β-effect, is [1] [7.4, 5]

$$\gamma \approx (2\pi u/\lambda)(1 - 2\pi\delta/\lambda) . \tag{7.2}$$

Evidently, for a smooth flow profile (weak pumping), the increment of instability is much smaller than for an abrupt profile (strong pumping) – other things being equal. That is, the instability is strong for an abrupt flow profile and weak for a smooth one. Our experiments indicate that a weak pumping of vortices by flows with a smooth profile is enough to maintain the stationary state of anticyclones but absolutely insufficient for cyclones.

In our opinion, substantiated in Chaps. 9–11, this asymmetry can be interpreted as follows. The large anticyclonic vortices under study are Rossby solitons. They are not prone to dispersion decay: their dispersion spreading is counterbalanced by their scalar nonlinearity. Their lifetimes are relatively long, limited mainly by viscosity. This is the reason why they survive with just a weak pumping which compensates for slight viscous losses of momentum. On the other hand, the cyclones are not Rossby solitons since their dispersion spreading is not counterbalanced by this nonlinearity. They undergo a rather rapid dispersion decay and their lifetimes are much shorter than those of the anticyclones. Thus they can survive only with a good amount of pumping supplied by a strong instability, with an increment close to that given by (7.1), and this requires an abrupt flow velocity profile (6.13). Such is our view on the physical basis of the cyclone-anticyclone asymmetry. In anticipation of the discussion to follow, let us observe here that, according to this view, the anticyclonic vortex JGRS, a Rossby soliton, is produced by counterflows with a smooth profile (6.12) while the cyclonic Jovian Brown Ovals, which are not solitons, are produced by flows with an abrupt velocity profile (6.13). We shall return to this point again.

The anticyclones shown in Figs. 7.1, 2a are traveling in rotating shallow water which, being a dispersive medium, permits the propagation of Rossby waves. The size of the anticyclones is much greater than r_R. The anticyclones drift in the direction of Rossby wave propagation at a speed close to V_R. They exist in the Rossby regime (2.3) and appear as large-amplitude geostrophic elevations with $h_0 \approx 0.5$. They carry along trapped particles of the liquid. These features of the vortices in question, especially the fact that large-scale ($a > r_R$) vortices

---

[1] To be more precise, on the threshold of instability the quantity $2\pi\delta/\lambda$ equals 0.8 rather than 1 as in (7.2); that is, the threshold wavelength is about an order of magnitude greater than the width of velocity shear [7.4].

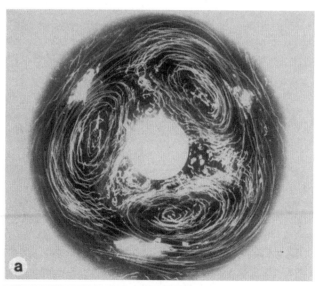

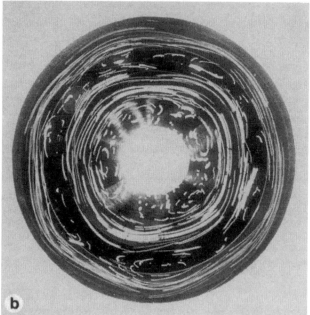

Fig. 7.2. Flow patterns in free-surface shallow water in the case of a smooth velocity profile **a)** for anticyclonic and **b)** for cyclonic vorticity of counterflows (experimental configuration as in Fig. 6.3)

are generated by flows in the regime (6.12), as well as their cyclone-anticyclone asymmetry, allow us to classify them as Rossby vortices.

Cyclones observed in our experiments also exist in the Rossby regime (2.3), appearing as large-amplitude geostrophic depressions with $a > r_R$. Therefore they are Rossby vortices, too. However, they are generated under conditions given by (6.13) which, in principle, can also give rise to vortices which bear no relation to Rossby vortices. An example of such vortices was observed in the experiments described in [7.6, 7] where the K-H instability was studied in concentric counterflows of a gas. The conditions of these experiments were designed so as to make the counterflows nearly symmetrical in the laboratory frame. The frame of rest for the disturbances generated by the instability is almost inertial, the Coriolis force can be disregarded and therefore the regime of the experiments is not the Rossby regime given by (2.3). Besides, the working gas can be treated as a practically incompressible medium under the conditions of [7.6, 7], which corresponds to $r_R \to \infty$ (see also [7.8]).

Also worth noting is the efficient generation of vortex chains in the Rossby regime (2.3) demonstrated in the experiments of [7.9, 10] which simulate terrestrial polar cyclones. Counterflows are produced here by the Coriolis force acting on the liquid pumped radially in a rotating circular channel. Stationary vortices are generated most efficiently when the liquid is pumped in near the inner and outer channel walls and drained through an axial slit in channel bottom (Fig. 7.3). It is easy to see that this flow pattern produces cyclones (Fig. 7.4). Indeed, under the action of the Coriolis force the liquid moving radially from the inner area to the slit (the common outlet) will slow down its rotation, lagging behind the vessel, while the liquid coming from the outer area will accelerate, rotating faster than the vessel. As a result, two longitudinal flows of opposite directions will arise in the rotating channel, the velocity curl in the transition area (that is, at channel axis, between the flows) directed along the vector of vessel angular rate, so the vorticity of the flow system will be cyclonic. If the flows are unstable, they will generate cyclones localized along the outlet slit – that is, along channel axis. Drawing the liquid together at the outlet – that is, at the centers of the vortices – will amplify the cyclones due to the conservation of angular momentum. In order to produce anticyclones, one should have the inlet ("fountains") rather than outlet ("funnels") at the channel axis. The liquid would then flow outwards from the centers of the anticyclones – that is, from the inlet – and the vortices would be damped rather than amplified. This is one of the reasons why the system described above is ineffective for generating anticyclones. There is practically no β-effect in the geometry used in [7.9, 10] (the vortices suffered almost no dispersion).

Since the experiments reported in [7.9, 10] are aimed at the investigation of cyclogenesis in terrestrial atmosphere, their distinctive features are:

(a) vortex size $a < r_R$;

(b) flow profile width $\delta < r_R$;

(c) no cyclone-anticyclone asymmetry as described above.

One can see that they differ radically in all respects from our experiments where the regime of the vortices corresponds to that of Rossby vortices on the giant planets as well as to that of drift vortices in magnetized plasma, satisfying $a > r_R, r_L$.

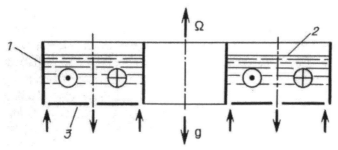

Fig. 7.3. Experimental configuration for studying the instability of sheared zonal flows produced in rotating shallow water by forced radial pumping: (1) circular channel rotating slowly about the vertical axis; (2) undisturbed free surface of the liquid; (3) bottom with three circular slits letting the liquid in and out of the vessel. The Coriolis force acting on the radial motion of liquid particles creates zonal counterflows in the channel (the indicated flow direction with respect to the vessel is obtained when the liquid, as shown by *arrows*, comes in through the slits at channel walls and is drained through the central slit which is the common outlet; the flow vorticity is cyclonic in the transition zone around the outlet). From [7.9, 10]

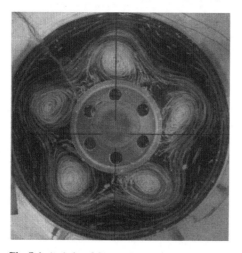

Fig. 7.4. A chain of five cyclones produced by unstable zonal counterflows. From [7.9, 10]

The experiments of [7.11, 12] should also be mentioned, where essentially the same vortex generation technique was used as in [7.9, 10]. They are discussed in detail in Sect. 7.5; here we only point out that they possess such features as

(1) generation of cyclones only (largely assisted, just as in [7.9, 10], by funnels created by the liquid draining into holes at the axis of channel bottom), and

(2) the absence of a free surface of the working liquid (water) which is sandwiched between two horizontal rigid lids, the upper lid being plane and the lower one (the bottom) somewhat inclined to create the β-effect; this geometry is equivalent to the approximation $c_0 \to \infty$ (a fluid incompressible not only in the three-dimensional but also in the two-dimensional sense) – that is, to $r_R \to \infty$ and $a \ll r_R$.

These experiments are thus inappropriate to the actual conditions on the giant planets where the large-scale, long-lived vortices are predominantly anticyclones and their sizes are larger than the Rossby radius.

In conclusion, we emphasize the fact that the technique based on forced radial pumping of the liquid can produce flows with large velocity gradients only if the gradients are of the cyclonic kind, which largely predetermines the asymmetry of the technique with respect to the sign of the vortices it generates: the conditions favor the production of cyclones. This important point can be better understood using as an example the experimental geometry of [7.9, 10] shown in Fig. 7.3. The Coriolis force imparts azimuthal acceleration to the liquid only in those areas where the liquid is moving in the radial direction. The azimuthal velocity of the liquid with respect to the vessel is therefore zero at the inlet; its magnitude rises smoothly towards the outlet near which it reaches its maximum. If the liquid is pumped in at both channel walls, then two oppositely directed longitudinal flows, each having the maximum possible velocity magnitude, meet in the transition area, near the common outlet at channel axis. Since this area is rather narrow, a velocity shear with an extremely high cyclonic vorticity is created there. Compared to such a strong velocity shear existing in this case near the outlet at the channel axis, the anticyclonic but very smooth shear occurring "on the way" to the outlet may evidently be neglected. Without this explanation, it could appear at first glance that counterflows with a strong anticyclonic vorticity could also be obtained in the system with radially pumped liquid we are discussing, by interchanging the inlet and the outlet. This is, however, not so. Indeed, if the inlet were placed at the channel axis and the outlets near the walls (Fig. 7.3), the longitudinal velocity imparted to the liquid by the Coriolis force would be nearly zero in the vicinity of the inlet and the velocity shear (now anticyclonic) between the azimuthal counterflows in the transition area would be insignificant.

We have discussed the technique based on radial pumping using the experiments of [7.9, 10] as an example; the situation is qualitatively the same in the experiments of [7.11, 12]. Hence the radial pumping technique is asymmetrical in principle: it is inappropriate for producing flows with large anticyclonic vorticities. The method used in our experiments is free from this disadvantage, allowing us to generate flows of high velocity shear with either sign of vorticity.

7.2 Self-Organizing Solitary Anticyclonic Rossby Vortex in Zonal Flows as a Model of Jovian Great Red Spot

The experiments carried out with the device shown in Fig. 6.4 yield results which are qualitatively similar to those described in the previous section, as far as the generation of chains of Rossby vortices is concerned. In particular, they indicate that the number m of vortices over the chain perimeter depends on the relative velocity of the counterflows: if the latter is low (a weak shear), up to ten anticyclones are observed; with increasing velocity the number of vortices in the chain becomes smaller. Figure 7.5 shows an example of the three-vortex mode.

Fig. 7.5. Flow pattern in free-surface shallow water for a smooth velocity profile and anticyclonic vorticity of the counterflows, experimental configuration as in Fig. 6.4.

A basically new result is that when the velocity of anticyclonic counterflows is sufficiently high, a structure with only one anticyclonic Rossby vortex on the perimeter of the paraboloid is produced [7.2, 3, 13–17], see Fig. 7.6. The vortex is large ($2a \approx (3\text{--}4)r_R$) and has a high amplitude ($h_0 \approx 1$); it drifts against the vessel rotation and efficiently entrains liquid particles (Fig. 7.7). The vortex is a self-organizing solitary structure in the counterflow system: as it grows, the flow profile changes drastically, adjusting itself to match the vortex (Fig. 7.7a). The profile is changed not only at the site of the vortex but also at the *opposite* side of the vessel.

Another important property of this solitary vortex is that its maximum vorticity is several times greater than that of the surrounding flows [7.14] – similar to what is observed in the case of large vortices in the atmospheres of the giant planets [7.18]. The solitary vortex thus produced may be considered stationary: it persists indefinitely long, although exhibiting slight deformations of an oscillatory type. Closed streamlines in the area where liquid particles have been entrapped by the solitary vortex are clearly seen in Fig. 7.6.

The photographs in Fig. 7.8 show consecutive drift phases of the solitary vortex with respect to the paraboloid. The vortex drifts against the vessel rotation at a steady rate close to the Rossby velocity.

Figure 7.9 illustrates the order of switching between vortex chains containing different numbers of vortices m as the counterflows' velocity is varied. The transitions between the instability modes corresponding to different values of m exhibit hysteresis which is an indication of the nonlinear nature of these phenomena. The highly distinct localization of vortices in the chain (Fig. 7.5) and of the solitary vortex (Fig. 7.6) also indicate a strong nonlinearity.

As we shall show in Chaps. 9–11, the solitary vortex in Fig. 7.6 may be considered a Rossby soliton. Because of its property to self-organize in unstable flows, we refer to it as an autosoliton. Accordingly, Fig. 7.5 demonstrates a chain of Rossby autosolitons.

In the generation of vortex chains (as in Fig. 7.5) and solitary vortices (as in Figs. 7.6, 8), a pronounced cyclone-anticyclone asymmetry is observed: as soon as the sign of the counterflow curl is reversed to cyclonic, all the structures shown in Figs. 7.5, 6, 8 cease to appear.

In order to identify the instability responsible for the generation of solitary Rossby vortices in the most interesting mode $m = 1$ (Fig. 7.6), one more important feature should be noted. In this mode, the outer ring is rotated counter to the vessel rotation at an angular rate approximately equal to $2\Omega_0$ (relative to the paraboloid). The velocity of the outer flow is then $u \approx \Omega_0 D \approx 300$ cm/s, and the corresponding Mach number is M$= u(g^* H_0)^{-1/2} \approx 7.5 > 2\sqrt{2}$ for $H_0 = 1$ cm. Consequently, the velocity jump across the discontinuity of the outer flow satisfies the criterion (4.2) for the stabilization of the K-H instability. Combined with the fact that cyclonic counterflows (obtained when the outer part of the vessel is rotated faster than its inner part) do not give rise to large vortices, this indicates that centrifugal instability may be responsible for the generation of the solitary vortex in this experiment. This is the instability which persists in a differentially rotating liquid even for M$\gg 1$ if the central part is rotated faster than the periphery. This instability does not arise if the counterflows' vorticity is cyclonic (Sect. 4.1). Therefore the solitary Rossby vortex discussed here (Fig. 7.6) is an example of the cyclone-anticyclone asymmetry in *the generation* of Rossby vortices. However, this example, as mentioned already, is related only to the supersonic flow regime with M$\gg 1$.

It must be stressed that the supersonic values of the Mach number in the experiment with the Rossby autosoliton are by no means necessary: they result

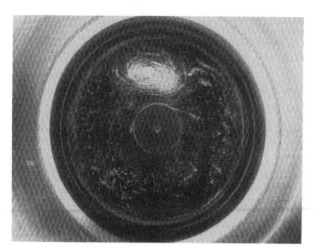

Fig. 7.6. A Rossby autosoliton in shallow water produced in the experimental configuration of Fig. 6.4 by counterflows of anticyclonic vorticity (the camera rotates with the vortex which drifts stationarily relative to the vessel in the clockwise direction – that is, against vessel rotation)

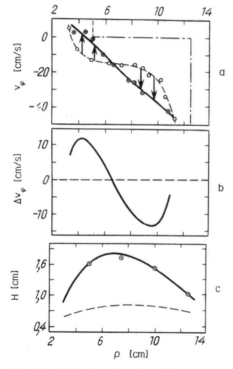

Fig. 7.7. (a) Meridional profiles of linear azimuthal velocity v_φ at liquid surface relative to the rotating paraboloid: no vortex (*dots and dashes*); in the vortex generation regime, within the vortex (*solid curve*) and in the region diametrically opposite to it (*dashed curve*). (b) Meridional profile of linear azimuthal velocity Δv_φ within the vortex, measured with respect to the flow, as shown by *arrows* in (a). (c) Meridional profiles of layer thickness H within the vortex (*solid curve*) and in the region diametrically opposite to it (*dashed curve*)

from the particular geometry chosen for the experiment. This is supported by the data obtained in the experiments with the three configurations described earlier, differing in counterflows' separation δ. These experiments show that the relative velocity of the counterflows $2u$ necessary to excite a vortex chain of a given mode increases with Ω_0, H_0, and δ. The flow velocity difference across the vortex ua/δ must ensure vortex rotation with a linear velocity significantly higher than its drift velocity $V_{dr} \approx V_R$: it is exactly then that a vortex with a clear-cut region of particle entrapment is observed (Fig. 7.6). The appropriate relation is

$$u \approx bV_R\delta/r_R \approx b\Omega_0^2\delta(H_0/g^*)^{1/2} \tag{7.3}$$

where b is a factor of the order of unity, the exact value depending on the details of the experiment. This implies, in particular, that it is the large value of δ used in the experiments on producing autosolitons that makes the Mach number $M = u(g^*H_0)^{-1/2}$ rather high.

This, however, creates no serious difficulties for the simulation of actual atmospheric vortices for which always M< 1. This is because, as indicated in Sect. 2.1, when dealing with natural vortices one must take into account the inhomogeneity of the medium along the vertical direction by replacing r_R with r_i. Then the Rossby velocity V_R is reduced by a factor of $(r_R/r_i)^2$ which amounts to dozens for real atmospheres. The Mach number corresponding to the velocity u obtained from (7.3) will be reduced by the same factor, hence the value of M will become quite close to the actual Mach numbers in planetary atmospheres.

Thus we conclude that large-scale Rossby vortices produced and studied in our experiments possess the main features of large-scale, long-lived natural vortices in the atmospheres of the giant planets. These features are: the size (in terms of r_R or r_i), the sign of vorticity, the direction and velocity of drift, the trapping of medium particles, the cyclone-anticyclone asymmetry, the stationarity, and the ability of self-organizing in unstable zonal counterflows. The experimental result that Rossby vortices are generated by unstable counterflows appears to be a physical model of the fact that the criterion (3.1) for two-dimensional (barotropic) instability of zonal flows is satisfied in the atmospheres of Jupiter and Saturn at the latitudes where the long-lived vortices are located.

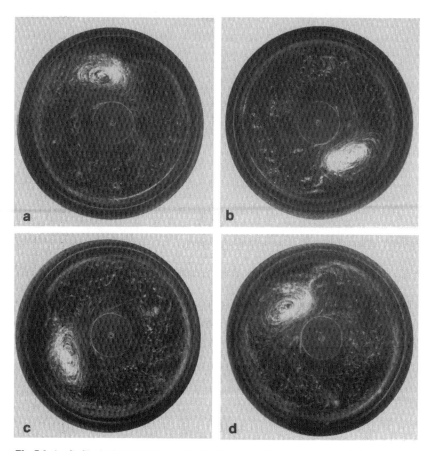

Fig. 7.8. (a–d) Clockwise drift phases of a Rossby autosoliton, counter to paraboloid rotation

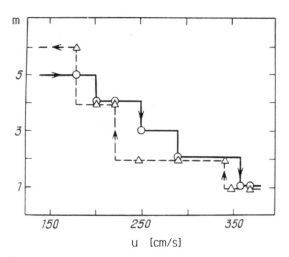

Fig. 7.9. Number of anticyclones in a chain (mode number m) as a function of the velocity of the ring producing the outer flow in the experimental configuration of Fig. 6.4 (*arrows* indicate the direction of velocity change)

7.3 Regularity in the Generation of Chains with Different Numbers of Vortices and the Problems of Uniqueness and Stationarity of the JGRS

The experimental results of the previous section suggest a simple interpretation for the intriguing uniqueness of the JGRS vortex over the entire Jovian perimeter. That is, they explain why another similar vortex cannot exist at a different site on the perimeter. In our opinion, the answer is that the JGRS is simply a sufficiently localized structure corresponding to the first mode of the hydrodynamic instability in zonal counterflows. It corresponds to the number of vortices over the perimeter $m = 1$ and develops under such counterflow conditions (velocity jump, shape of velocity profile) which preclude the existence of the second and higher order modes ($m \geq 2$), that is, of vortex chains corresponding to smaller instability wavelengths.

In order to envision the probable mechanism of mode competition and switchover between the modes with different numbers of vortices in the chains, let us turn to the expression (7.2) for the growth rate of the K-H hydrodynamic instability in flows with a velocity shear. Although this expression does not take into account liquid compressibility, the β-effect, and the centrifugal effect, we nevertheless believe it to be qualitatively adequate for understanding the phenomena in question. It is natural to assume that the instability will exist if its growth rate (increasing with flow velocity) is at least greater than the decrement of vortex viscous decay $1/\tau_\nu$ where τ_ν is the characteristic viscous time given by (5.35) which is of the same order of magnitude as the Eckman time τ_E given by (4.8). We then obtain the following instability threshold, i.e. the minimum velocity of the counterflows at which the instability sets in:

$$u_c \approx \frac{(\nu \Omega_0)^{1/2}}{2\pi H_0} \frac{\lambda^2}{\lambda - 2\pi\delta} \tag{7.4}$$

where $\lambda = 2\pi\varrho_0/m$ is the wavelength and ϱ_0 is the radius of the parallel which marks the borderline between the counterflows.

It can be seen from (7.4) that the instability arises only when $\lambda > 2\pi\delta$, which is well known [7.4, 5]. The minimum value of the threshold velocity

$$(u_c)_{\min} \approx \frac{8\delta}{\tau_E} \tag{7.5}$$

corresponds to the instability mode with wavelength $\lambda \approx 4\pi\delta$, that is, with mode number

$$m \approx \frac{\varrho_0}{2\delta} . \tag{7.6}$$

The instability growth rate (7.2) reaches its maximum also at

$$\lambda \approx 4\pi\delta , \tag{7.7}$$

and this maximum is

$$\gamma_{max} \approx \frac{u}{4\pi\delta} \ . \tag{7.8}$$

The following nonlinear effect (supposed by us) must be emphasized here: the value of δ increases with increasing amplitude of the oscillations and with decreasing mode number m. In other words, the larger the mode (the smaller m), the stronger it broadens the initial flow profile (Fig. 7.7 may serve as an example).

Let us now assume that during the formation of counterflows their separation δ has some initial value and their velocity is increasing gradually, starting from zero – like in the experiment leading to transitions between longitudinal modes illustrated in Fig. 7.9. After the minimum threshold given by (7.5) is exceeded, an instability of mode m at wavelength $\lambda \approx 4\pi\delta$ sets in, since this is the wavelength corresponding to the minimum instability threshold. Now let the counterflow velocity u keep increasing. The oscillations will then develop further and, according to what has been said above, the value of δ will rise, shifting the growth rate maximum (7.8) towards longer wavelengths. After a certain velocity value u is reached, the larger mode with the next smaller longitudinal number $m - 1$ becomes more favored. This larger mode, as it starts developing, will further broaden the velocity profile – that is, it will increase δ. As a result, the growth rate of mode m will fall and this mode will cease to be generated. Mode transition will thus occur, $m \rightarrow m - 1$, mode $m - 1$ emerging near its excitation threshold. Similarly, as the velocity u keeps increasing, mode $m - 2$ will arise in place of mode $m - 1$, and so on. Such is evidently the mechanism of nonlinear competition between longitudinal modes, underlying the single-step transitions between the modes as illustrated in Fig. 7.9. It is easy to see that if the direction of flow velocity change were reversed, mode transitions should exhibit hysteresis, which is exactly what is observed in the experiments (Fig. 7.9).

In the laboratory, the value of flow velocity shear is varied manually. The analogy with the cyclogenesis conditions in Jovian atmosphere is due to the fact that, according to modern concepts, the zonal flows result from the merging of vortices. This two-dimensional turbulent cascade develops gradually in time, thus causing a gradual increase in flow velocity shear, which is basically the same as what is done in the experiments.

Figure 7.10 illustrates the gradual increase in the velocity of zonal flows created by the merging of vortices in a numerical simulation of upper Jovian atmosphere [7.19–22].

The evolution of the JGRS vortex may thus be conceived as follows. Formerly, at an early stage of the turbulent cascade, when the counterflows' velocity shear at the Jovian latitude in question (22°S) was relatively low, a chain of rather small vortices was created at that latitude. As the cascade developed, the velocity shear became higher and the number of vortices in the chain decreased. The newly arising structures were larger than their predecessors, broadening the flow profile still further. Eventually, the development of the cascade reached its contemporary phase where the most favorable longitudinal mode suppressing all the

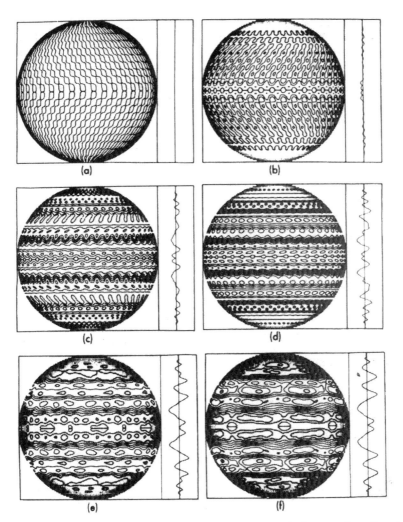

Fig. 7.10. Zonal flows developing from small-scale turbulence in the atmosphere of a rotating planet (barotropic model for the conditions of Jupiter): (**a**) initial perturbation of the stream function; (**b–f**) temporal evolution of the perturbation (meridional profiles of averaged velocity directed along the parallel are shown to the right of each picture). From [7.20]

other (smaller) ones is the mode with $m = 1$. This corresponds to the present-day JGRS vortex.

In this way, the question: "What is preventing another such vortex from arising at some other location on the same Jovian parallel?" could be answered as follows: "The existence of mode $m = 1$ which broadens the zonal flow profile to an extent that precludes the generation of mode $m = 2$".

The relevant quantitative assessment can proceed in the following manner. It would be logical to assume the width of the counterflows' profile at the JGRS site to be close to the width of the vortex itself. Counting only the particle capture region, the latter width is $1.3 \cdot 10^4$ km [7.23, 24]. Thus $\delta \approx 2 \cdot 10^4$ km may be assumed. This value of δ appears to be realistic also because it is close to the Rhines length given by (2.19). According to (7.6, 7), the most favorable wavelength for the generation of vortices is then $\lambda \approx 4\pi\delta \approx 2.5 \cdot 10^5$ km which is somewhat greater than half the perimeter of the JGRS parallel (the planet equator is $4 \cdot 10^5$ km). This means that there is no room for two vortices separated by a distance of λ at this parallel. In addition, it should be kept in mind that the actual structure of the zonal flows, oscillatory along the meridian, is much more complicated than the flow we have assumed here for our model.

With all that has been said, the fact that the JGRS is unique over the entire planet perimeter should no longer appear surprising. According to the above arguments, the zonal flow velocity profile at the JGRS latitude is too wide to allow more vortices than one to be generated. At other parallels, the flow profile is narrower, permitting, in particular, the Jovian chains of White and Brown Ovals to be generated.

Now let us consider the problem of JGRS stationarity. Recall that, according to the data obtained from the Voyager space missions, the JGRS vortex in the upper atmosphere of Jupiter is an anticyclone with a relative amplitude $h_0 \approx 0.1$, floating along the parallel in the more or less isothermal cloud layer with an effective thickness $H_0 \approx 25$ km [7.24–28]. (The latter value corresponds to atmospheric density decrease by a factor of e along the vertical; the actual layer thickness may be several times as large.) This layer is physically distinguished by its temperature which is the lowest over the entire height range – about 130 K (the average pressure here is about one-third of an atmosphere). Beneath this cloud layer, beginning at the level where the pressure is about half an atmosphere, lies a more extensive layer about 1000 km thick [7.28, 29], resting upon a liquid hydrogen support (Sect. 6.2).

We are now going to show that in order to explain the fact of stationarity of the JGRS, which has been observed for over 300 years, one has to assume that the vortex is being pumped by the (unstable) zonal flows, since otherwise it would have long ceased to exist due to viscosity effects. The characteristic time of viscous damping for the JGRS vortex can be estimated on the basis of (5.35) with r_R replaced by r_i: $\tau_\nu \approx 10^2 \tau_E \approx 10^2 H_0 (2\nu\Omega)^{-1/2}$ for $a \approx 10r_i$ where $\Omega \approx 10^{-4}$ rad/s. For the estimate of τ_ν, the kinematic viscosity can be assumed to be $\nu \approx 10^5$ cm^{2}s^{-1}, a value five orders of magnitude higher than hydrogen molecular viscosity and generally accepted for planetary atmospheres [7.30, 31]. We then obtain $\tau_\nu \approx 3$ years, whereas the JGRS is observed for a period more than two orders of magnitude longer than that.

It is worth noting that the above estimate for τ_ν, obtained from (5.35) with the replacement $r_R \to r_i$ and with $a \gg r_i$, is practically independent of H_0 since $a^2/r_i^2 \propto 1/H_0$. In particular, even if one takes $H_0 \approx 1000$ km (the entire thickness down to the liquid hydrogen layer), the value of τ_ν will increase less than twofold, which will not change the general character of the situation.

Taking these arguments into consideration, the above assumption that the observed JGRS stationarity is due to an intensive pumping of the vortex by the (unstable) zonal flows appears quite realistic. As regards the detailed mechanism of vortex generation by unstable flows with a velocity shear, it is the mechanism that generates Kelvin's "cat's eyes" [7.32, 33]. The JGRS is accordingly a solitary "cat's eye".

We believe that the arguments presented here solve in principle the problem of uniqueness and stationarity of the JGRS. On the other hand, they are only qualitative, of course, and need to be backed with nonlinear analysis and numerical simulation.

7.4 Two- and Three-Dimensional Models of the JGRS

Thus it appears that one may consider the experimental results described here as an instance of a successful laboratory simulation of the long-lived Rossby vortices in the atmospheres of the giant planets and their generation mechanism. So far, however, this simulation has been carried out only at the qualitative level, that is, in a purely two-dimensional medium, the vertically homogeneous shallow water. Real planetary vortices are substantially inhomogeneous along the vertical, and if this is not taken into account, a drastic quantitative discrepancy between the two-dimensional laboratory model and its natural prototype may result. In order to elucidate this problem, we are going to do the following. First, assuming the JGRS to be a two-dimensional (barotropic) Rossby vortex, we shall estimate its principal parameters (the expected radius a, the drift velocity V_{dr}, the characteristic velocities of proper rotation – linear v_0 and angular $\omega_0 \approx v_0/a$, the ratio $\varepsilon = v_0/V_{dr}$). After that, we shall compare these parameters to the observations and to the results yielded by the three-dimensional model, differing from the two-dimensional one in that r_R is replaced by r_i. Here are these comparisons [7.34, 35].

Recall that the main parameters of Jovian atmosphere at the JGRS latitude (22°S) have the following values: $f \approx 1.2 \cdot 10^{-4}$ s^{-1}, $\beta \approx 4.6 \cdot 10^{-14}$ s^{-1}cm^{-1}, $g \approx 2.5 \cdot 10^3$ cm s^{-2}, $H_0 \approx 2.5 \cdot 10^6$ cm, $r_R \approx 6 \cdot 10^8$ cm; the observed amplitude of atmosphere elevation in the vortex is $(\delta H)_0 = h_0 H_0 \approx 0.1 H_0$ [7.27–29]. These data make it possible to determine what would be the values of the JGRS parameters if it were a two-dimensional smooth Rossby soliton satisfying (5.29, 30). These values, calculated with the aid of (2.11) and (5.9), are given in the upper row of Table 7.1.

Let us also recall that, as indicated already in Sect. 2.1, the baroclinic Rossby radius $r_i \approx r_R/6$ rather than the Rossby-Obukhov radius r_R ought to be used as the characteristic dispersion scale for a vertically inhomogeneous atmosphere. (The factor of 1/6 in (2.10) was obtained under the assumption that the upper Jovian atmosphere is isothermal. This is not exactly so but quite adequate for the purpose of our illustrative estimation.) Therefore, if a three-dimensional (baroclinic) Rossby soliton exists, there are good grounds to believe that, for the most part, the three-dimensional solitonic model of natural vortices like the JGRS differs from the two-dimensional model merely in that r_R is replaced with r_i. (For definiteness, we mean here the model based on the smooth Rossby soliton.) Replacing r_R with r_i, we obtain the second row of Table 7.1 [7.2, 34, 35]. The third row lists the observational data obtained during the Voyager space missions. The observations indicate (see, in particular, [7.23, 24, 36–38] and the references cited there) that the JGRS vortex carries along captured particles of the medium – in accordance with its large nonlinearity indicator $\varepsilon \approx 20$ (see also [7.35]). The three-dimensional solitonic model yields a close enough value of nonlinearity $\varepsilon \approx 10$, while the two-dimensional model gives $\varepsilon \ll 1$, contradicting the observations not only in the quantitative but also in the qualitative sense, because such a low value of the nonlinearity parameter ε is incompatible with the existence of a particle capture region in the JGRS. It is seen from Table 7.1 that the observational results disagree substantially with the two-dimensional model (which, for instance, overestimates the soliton drift velocity by more than a factor of 50). On the other hand, the observed values are close enough to the data of even the three-dimensional model which is based on the smooth Rossby soliton. Evidently the agreement will be still better if the requirement of soliton smoothness is dropped – that is, if the conditions of (5.29, 30) are not imposed on soliton parameters.

Table 7.1 Comparison of the two- and three-dimensional solitonic models of the JGRS with the observations

	$2a, 10^3$ km	V_{dr}, m/s	ω_0/f	$\varepsilon = v_0/V_{dr}$
Two-dimensional model	66	160	$3 \cdot 10^{-3}$	$7 \cdot 10^{-2}$
Three-dimensional model	11	4.5	$11 \cdot 10^{-2}$	10
Observations	12*	3	$9 \cdot 10^{-2}$	20

*This is the size of the particle capture region (measured along the meridian). The actual value of $2a$ may be about twice smaller, that is, about $6 \cdot 10^3$ km

Since the three-dimensional model, in fact, differs from the two-dimensional one only in the quantitave detail ($r_R \rightarrow r_i$), it may be termed quasi-two-dimensional. And the qualitative aspects are fairly well simulated in the two-dimensional experimental configuration. When the experimental results are corrected by substituting r_i for r_R, quantitative agreement with the simulated natural object is achieved as well. These findings indicate, in particular, that the JGRS vortex may be viewed as a scalar monopolar Rossby soliton. This view is also supported by the fact that, as can be easily seen, the time span during which the JGRS vortex is being observed is about five thousand times longer than the dispersion spreading time of a linear baroclinic Rossby wave packet of the same size (Table 2.1, Fig. 5.3).

We should mention another common way of taking into account the three-dimensionality of the natural atmospheric vortices in question. The vortex under study (in this case, the JGRS) is assumed to be floating in a thin upper cloud layer which rests on a much more thick and dense lower layer; the densities of the two layers are assumed to be ϱ_1 and $\varrho_2 = \varrho_1 + \Delta\varrho$, respectively. The baroclinic Rossby radius, by analogy with (2.9), is

$$r_i = \frac{(g'H_1)^{1/2}}{\pi f} \tag{7.9}$$

where H_1 is the thickness of the upper layer and the reduced acceleration of gravity $g' = g\Delta\varrho/\varrho_2$ is a free (adjustable) parameter which is chosen so as to achieve the best agreement between the model and the observations. The uncertainty in g' is due to the uncertainties in ϱ_1 and ϱ_2 which vary significantly with depth. Choosing $g'/g = 1/2\text{–}1/4$ in accordance with [7.22, 39, 40], we obtain from (7.9): $r_i/r_R = 1/4 - 1/6$, which is virtually the same as the value of r_i/r_R defined by (2.10) above in a more consistent fashion.

The laboratory simulation results discussed here also appear to provide an interpretation for some of the general effects observed in the oceans. For example, the pronounced cyclone-anticyclone asymmetry of intrathermoclinic vortices is apparently associated not only with the mechanism of their genesis but also with the different nonlinear-dispersion behavior of the vortices of opposite polarity. In other words, this asymmetry may be due to the same mechanism as the similar phenomenon in the atmospheres of the giant planets, which has been interpreted in the shallow-water simulations described here. This viewpoint, substantiated in detail in a theoretical study [7.41], will be discussed in Sect. 10.3.

7.5 Alternative JGRS Laboratory Models

There are two other methods used in laboratory investigation of geostrophic vortices in the Rossby regime [7.11, 12, 42, 43]. We shall now discuss the principal results obtained with these methods.

In the experiments described in [7.42, 43], liquid is enclosed in a ring channel formed by two cylinders and a horizontal bottom, rotating all together about the common vertical axis. A controllable radial gradient of temperature is maintained in the liquid, giving rise to a small gradient of density. The latter gradient, being noncollinear with gravity, produces an azimuthal flow in the liquid – the so-called "thermal wind" [7.30]. (This phenomenon is similar to plasma drift in crossed electric and magnetic fields.) The flow velocity varies along the vertical, making the flow structure three-dimensional. If somewhere across the channel the temperature and density gradients change their signs, counterflows arise. The experiments [7.42, 43] indicate that under certain conditions the flows become unstable in this configuration, producing vortex chains of different azimuthal mode numbers. In particular, the conditions can be adjusted so that, with liquid density at its minimum (the temperature at its maximum) midway across the channel, the mode $m = 1$ sets in, where only one anticyclone (preceded, however, by a weaker cyclone) fits in over the entire system perimeter. This vortical structure is proposed by the authors of [7.42, 43] as a model for the JGRS. They have not been able to create a density extremum of the opposite sign (a maximum in the middle of the channel) and simulate this situation numerically. This calculation yields a cyclonic structure resembling the Brown Ovals of Jupiter. As far as the applicability to the JGRS is concerned, this model, which is based on thermogyroconvection in deep water, differs quite significantly from our model, based on solitary Rossby vortices in shallow water. There are four principal discrepancies between the model described in [7.42, 43] and its natural counterpart:

1) The horizontal vortex dimensions in this model are much smaller than the depth of the liquid (deep water). On Jupiter, however, the horizontal size of the JGRS is much greater than the effective thickness of the atmosphere. 2) The model does not explain the cyclone-anticyclone asymmetry observed in nature. 3) In order to explain the properties of the JGRS vortex on the basis of this model, one has to assume a temperature maximum (density minimum) at its center, which qualitatively contradicts the observational data [7.25, 27]. 4) The horizontal dimensions of the model vortex are much smaller than the Rossby-Obukhov radius and are approximately equal to the baroclinic (inner) Rossby radius r_i. This feature makes the vortex in question similar to the relatively small Jovian vortices but dissimilar to the JGRS which has $a > r_R$.

Nevertheless, there is some qualitative similarity between the three-dimensional model presented in [7.42, 43] and our two-dimensional model discussed above. First, in both cases the self-supporting vortical structure results from an instability of zonal counterflows (although the latter are created in different ways). Second, the vortex sizes are physically comparable: they are determined by the Rossby scale which is the "two-dimentional" radius r_R for the two-dimensional model and the "three-dimensional" radius r_i for the three-dimensional one.

The second alternative model of natural atmospheric vortices [7.11, 12] is produced in the experimental configuration shown schematically in Fig. 7.11. The setup is rather similar to that shown in Fig. 7.3, differing in some details. The maximum angular rate of rotation is about 4 rps. The working liquid (water) is contained between two coaxial vertical cylinders and two solid lids of which the upper is horizontal and the lower (the bottom) slightly inclined so that there is a gradient of liquid layer thickness directed towards the periphery, producing the same β-effect as in (5.9), see also Supplements S5.2, 3. There are six groups of holes in the bottom, equally spaced about the azimuth, two holes in each. Water is pumped in through the inner holes and drained out through the outer ones. The radial current produced in this way is affected by the Coriolis force, creating in the vessel a sheared flow directed counter to the overall system rotation. The rotation rate of the vessel and the flow velocity correspond to the Rossby regime. The maxima of flow velocity gradient and flow vorticity are in the middle part of the ring channel. A flow with such a velocity shear is unstable, generating a chain of vortices drifting with respect to the vessel frame at the local flow velocity. The vortex generation mechanism is similar to that of the experiments [7.9, 10] discussed earlier.

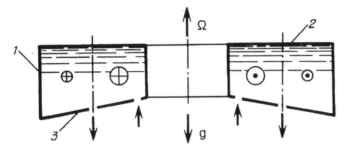

Fig. 7.11. Schematic of the experiment for producing cyclones by a sheared flow created in a rotating liquid by forced radial pumping: (**1**) circular channel filled with the liquid, rotating about the vertical axis; (**2**) rigid lid; (**3**) bottom inclined in order to obtain β-effect, with six pairs of holes distributed uniformly over the azimuth. As shown by *arrows*, the liquid is pumped into the channel through the holes at its inner wall (the inlets) and is drained through those at its centerline (the outlets). The Coriolis force acting on radial particle motion creates a flow directed against the general rotation in the vessel's frame. Flow velocity is higher at the inner wall than at the outer one (the pictograms in the figure are accordingly of different size); at channel axis, over the outlets, the vorticity is cyclonic and its magnitude is the highest

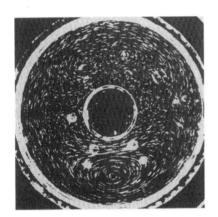

Fig. 7.12. A cyclonic vortex generated by unstable sheared flows. From [7.11, 12]

The strongest shear, where the gradient of flow velocity is the highest, is cyclonic (see the explanation of the experiments [7.9, 10] in Sect. 7.1). Accordingly, the vortices created in this way are cyclones, just as in [7.9, 10] (see Fig. 3 in [7.11]). Anticyclones, as pointed out in Sect. 7.1, cannot be effectively generated in such systems.

A new feature as compared to [7.9, 10] is that when the flow velocity is sufficiently high, that is, when the working liquid is pumped intensively enough, the number of cyclones in the chain can be reduced to only one over the entire perimeter – Fig. 7.12. The vorticity of the structure shown in Fig. 7.12 has its maximum in the center of the vortex.

The experiments of [7.11, 12] indicate that only one type of intense and stable vortices, the *cyclones*, can be generated in the way described above. It is impossible to create intense and stable anticyclones (by reversing the radial pumping direction): anticyclones, if any, live no longer than one or two proper revolutions.

The authors of [7.11, 12] state to have produced a laboratory simulation of the JGRS, supporting their conclusion with the following arguments: 1) the existence of a regime with a single vortex over the system perimeter; 2) the merging of vortices which occurs when one azimuthal instability mode is replaced by another as the sheared flow velocity increases; this process resembles the merging of vortices in the upper atmosphere of Jupiter [7.23, 24, 38]; 3) the localization of the vortices in the region of maximum horizontal velocity shear; 4) the fact that vortex length-to-width ratio is about 2:1; 5) the vorticities of the vortex and of the flows which generate it are of the same sign (this is evidently trivial).

The experiments in question have the following two specific features. First, the intensive vortices generated in these experiments have only the cyclonic polarity. This precludes the simulation of the cyclone-anticyclone asymmetry which seems to be a fundamental property of the Rossby vortices in the atmospheres of the giant planets.

Second, the presence of a solid upper lid in the experiments of [7.11, 12] implies that the liquid has no free surface and is therefore incompressible not only in the three-dimensional but also in the two-dimensional sense. This corresponds to the limiting case of $c_0 \to \infty$, that is, $r_R, r_i \to \infty$, hence the vortex size is $a \ll r_R, r_i$ – in contrast with the condition $a > r_R$ or $a > r_i$ which holds for the JGRS.[2] On the other hand, these experiments can be viewed as a simulation of *cyclonic* vortices such as the Jovian "Barges".

From the physical viewpoint, attention should also be drawn to the following interesting results of the experiments described in [7.11, 12]. The number of vortices in the chain becomes smaller as the flow velocity shear increases. This tendency is qualitatively similar to what has been observed in our experiments (Sect. 7.3), despite the totally different geometry of the latter (shallow water, free surface, large vortices with $a > r_R$, etc.). This is an additional demonstration of the common character of instabilities in flows with velocity shear.

The authors of [7.11, 12] also point out that in the presence of small-scale turbulence, often encountered in actual laboratory experiments, the merging of

---

[2] This calls for an explanation of what is meant under "vortex size" a. If the visible size of the JGRS is meant or the size corresponding to the maximum linear velocity of rotation in the vortex, then $a > r_R$. If, on the other hand, a is understood as the characteristic size corresponding to the gradient of rotation velocity, then $a \lesssim r_R$, but $a > r_i$ – see [7.44, 45] on this subject.

vortices proceeds much faster than in the viscous time. We shall take this into account in Sects. 9.7 and 10.2.

Another interesting result by the same group of authors is their experimental simulation of oceanic flows such as the Gulf Stream and the Kuroshio [7.46].

7.6 Stationary Rossby Vortices in Flows and the Blocking Phenomenon

The study of Rossby vortex propagation in planetary atmospheres in the presence of wind reveals some interesting and nontrivial features which we are going to illustrate now. Let a Rossby wave be propagating in a stream of gas which is moving at velocity u (the positive direction is assumed to be eastwards). If for $u = 0$ (no wind) the atmosphere had a constant thickness ($H_0 = $ const), then a value of $u \neq 0$ will give rise to a meridional thickness gradient in the atmosphere, balancing out the Coriolis force acting on the wind:

$$g^* \frac{\partial H_0}{\partial y} = -fu \ . \tag{7.10}$$

The Rossby wave velocity in the planet's frame V will now be different from its initial value (in the absence of wind) due to two effects: first, because the wave is carried along by the wind, and second, because its drift velocity with respect to the medium – that is, to the wind – has changed according to (5.11) since a gradient of H_0 has appeared. Consequently,

$$V = \frac{\omega}{k_x} + u = -\frac{r_R^2 \beta(u)}{k^2 r_R^2 + 1} + u \tag{7.11}$$

where, according to (7.10) and (5.11),

$$\beta(u) = -\frac{f^2}{H_0} \frac{\partial}{\partial y} \frac{H_0}{f} = \frac{\partial f}{\partial y} - \frac{f}{H_0} \frac{\partial H_0}{\partial y} \ ,$$

or

$$\beta(u) = \beta(0) + \frac{u}{r_R^2} \ , \tag{7.12}$$

that is, the Rossby velocity with respect to the wind:

$$V_R(u) = V_R(0) + u \tag{7.13}$$

where $V_R(0) = \beta(0) r_R^2$ is the Rossby velocity in a uniform atmosphere in the absence of wind. The relation (7.13) reflects an interesting fact: an eastward wind of velocity u increases the westward Rossby velocity with respect to the wind exactly by u (with the sign of u duly taken into account). If, for example, the wind is blowing to the east, then $u > 0$ and the Rossby velocity relative to the wind becomes greater; if it is blowing to the west, $u < 0$ and the relative Rossby velocity decreases. Therefore, should a wave propagate at the Rossby velocity, the velocity of the wave in the planet's frame (7.11) would be independent of wind velocity, according to (7.13). But a Rossby soliton does propagate at nearly the Rossby velocity. Consequently, Rossby solitons cannot be carried off by winds at all.

In the general case, the velocity of a Rossby wave is not equal to the Rossby velocity but depends on the wavelength. From (7.11, 12) we find the wave velocity in the planet's frame to be

$$V = \frac{\omega}{k_x} + u = \frac{k^2 r_R^2 u - V_R(0)}{k^2 r_R^2 + 1}.$$

And since, according to (5.11), $V_R(0) = r_R^2 \partial f / \partial y$, then, as long as $k^2 u > \partial f / \partial y$, the Rossby wave which had been traveling westwards in the absence of the wind will be carried eastwards by it (for more details see also [7.30]). If

$$k^2 u = \frac{\partial f}{\partial y} , \tag{7.14}$$

the wave will be stopped by the wind ($V = 0$). If such a wave has a sufficiently large amplitude, an effect interesting for meteorology may arise. Namely, if wave amplitude complies with the particle capture condition (2.15), the wave will possess the properties of a "genuine" vortex: it will keep its own particles and let no foreign particles in. When such a strongly nonlinear "standing" Rossby wave has been formed in the atmosphere, the following consequences may happen. After this stagnant Rossby vortex runs out of its "own" precipitation, a draught may occur in the area occupied by it, similar to some known natural phenomena such as the 1972 draught in the USSR. These are the so-called blockings, currently under study by a number of researchers [7.47, 48]. It is seen from (7.14) that a blocking can arise, for instance, when a Rossby wave of wavelength $\lambda = 2\pi/k \approx$ 3000 km is traveling up an eastward wind of several meters per second.

It is important to note that these effects depend on the fact that $\beta(u) \neq \beta(0)$. This inequality holds (for $u \neq 0$) only if the "shallow water" has a free surface: with no free surface, shallow water incompressibility leads to the limiting case of $c_0 \rightarrow \infty$ and $r_R \rightarrow \infty$, implying $\beta(u) \rightarrow \beta(0)$ which cancels the effects we are discussing.

Here are some further examples illustrating the influence of the phenomena discussed above on the effects depending on $\beta(u)$. One example is a dipolar Rossby vortex drifting at $V_{dr} = u$ (Sect. 5.5). The profile of such a vortex [7.49] depends on its shielding length $\mu = [\beta(V_{dr})/V_{dr}]^{1/2}$, whence with the aid of (7.12) we obtain $\mu = (1/r_R^2 + \beta/|u|)^{1/2}$ (the latter relation to be compared with (5.24)). Therefore, even in the absence of an "initial" β-effect, a dipolar vortex in a free-surface shallow water will be spatially shielded within a length close to r_R.

A second example is the Rayleigh-Kuo criterion (3.1) for zonal flow instability, taken in the "rigid lid" approximation where the free surface of shallow water is disregarded. If the liquid has a free surface then, according to (7.12), $\beta(0) \to \beta(0) + u/r_R^2$ and the criterion (3.1) is modified as follows [7.2, 35, 50, 51]:

$$u'' - \beta(0) - \frac{u}{r_R^2} = 0 \ . \tag{7.15}$$

This is more accurate than the approximate criterion given by (3.1). One can see that the instability criterion of (7.15) is more easily satisfied if there is a westward mean flow ($\langle u \rangle < 0$). This effect must be kept in mind when dealing with problems related to the stability of zonal flows [7.26, 52].

One interesting fact is worth mentioning. Equations (7.12, 13) imply that the dispersion of a Rossby wave (or vortex) in a westward flow ($u < 0$) faster than the Rossby velocity ($|u| > V_R(0)$) changes its sign ($\beta(u) < 0$). This means that while in the case of $|u| < V_R(0)$ a balance between dispersion and nonlinearity (giving rise to a Rossby soliton) is possible only for anticyclones, a similar balance in the case of $|u| > V_R(0)$ is possible only for cyclones. The latter situation is sometimes simulated in numerical calculation [7.22, 39, 40, 53], but we are unaware of any examples for it to occur in the nature. As regards flow stability problems in general, the reader is referred to the reviews in [7.32, 33, 54–56].

The following observational facts, similar to the blocking phenomenon in their nature, should also be pointed out. The ozone distribution in the circumpolar terrestrial atmosphere was found to be in a qualitative agreement with potential vorticity isolines of the polar Rossby vortex. Latitudinal ozone displacements are observed only when the polar vortex is either absent or decaying due to its instability. See [7.57–60] and the literature cited there for more on this problem area.

The principal contents of this chapter may thus be summed up as follows. Laboratory simulation of Rossby vortex generation by unstable sheared flows similar to the zonal flows in planetary atmospheres has been carried out. Efficient vortex generation has been demonstrated.

A pronounced cyclone-anticyclone asymmetry in generating stationary vortices of different polarity has been discovered: stationary anticyclones are easy to create, even with a very smooth velocity shear; on the contrary, cyclones can be generated only with an extremely sharp shear, that is, with as high flow velocity gradient as possible. The rules discovered in the laboratory simulations are in good agreement with the cyclone-anticyclone asymmetry phenomenon observed in the atmospheres of the giant planets, whose large long-lived Rossby vortices are almost invariably anticyclonic, cyclones being an almost unique exception. The proposed interpretation of this phenomenon is that large ($a > r_i, r_R$) Rossby anticyclones are solitons, while the cyclones are not and therefore experience a relatively fast dispersion decay, leading to relatively small lifetimes.

A fundamental rule has been discovered: when vortex chains are generated by unstable circular sheared flows, the number of vortices in the chain decreases with increasing velocity shear. For a strong enough velocity shear, only one large-scale anticyclonic vortex remains in the chain – a Rossby autosoliton, self-organizing in the flows and drifting steadily counter to the system rotation. This vortex is a stationary laboratory model of the natural vortices such as the JGRS. The fact that the JGRS is unique over the entire planet perimeter has been interpreted.

Laboratory simulation has been shown to yield both qualitative and quantitative agreement with astrophysical observations. In order to achieve this, it is necessary and sufficient to use the replacement $r_R \rightarrow r_i$, thus making a transition from the two-dimensional (barotropic) model to the three-dimensional or, more precisely, quasi-two-dimensional (baroclinic) one.

It has been shown that laboratory simulation of atmospheric Rossby vortices is adequate only if the rotating shallow water has a free surface. In this case, the both most conspicuous features of the natural Rossby vortices on the giant planets can be simulated, namely, their cyclone-anticyclone asymmetry and their large scale: $a > r_i$ (and even $a > r_R$ which is valid for the JGRS).

The hypothesis has been proposed that the cyclone-anticyclone asymmetry observed in the long-lived intrathermoclinic vortices ("lenses") in the oceans might have the same physical nature as that of the large atmospheric vortices on Jupiter. A further substantiation of this concept is given in Chap. 10.

Some general rules observed in various experiments with sheared flows which are greatly different in their parameters draw one's attention. For instance, despite the obvious distinction between the experiments described in [7.11, 12] and in [7.13–17, 61], a similar result has been observed in both cases: the number of vortices in the chain m decreases with increasing counterflows' velocity u. The same has been also observed in the experiments of [7.6, 7] despite the very great dissimilarity of their experimental conditions from those of [7.11, 12], namely, the Reynolds number being two orders of magnitude less, the Eckman number seven orders of magnitude higher (and larger than unity), and the different scenaria of the transition processes in the switchovers between chains with different numbers of vortices. Thus, in [7.11, 12] the transition $m = 4 \rightarrow m = 3$ is accompanied by oscillations which displace the vortices in the radial direction. The vortices therefore are shifted to areas with different flow velocities; as a result, adjacent vortices are drawn by the flow at different velocities, coming close to each other and merging. In [7.6, 7], the transition process is of another type: vortex size oscillations are developing, the larger vortex absorbing the nearby smaller one.

So the falling character of the $m(u)$ dependence appears quite fundamental. To forestall, we note that a similar rule is also observed in our other experiments, described in the next chapter.

In the connection with what have been said above, a recent study [7.62] is worth mentioning which gives the theory of mode switchovers in a circular flow with a velocity shear. This theory is essentially based on arguments which are very close to those discussed in the present chapter.

8. Laboratory Simulation
of Galactic Spiral Structures

The present series of experiments is, from the hydrodynamic viewpoint, a logical sequel to the experiments described in the previous chapter where differentially rotating shallow water was considered a model of an ocean or atmosphere. In this chapter, it will serve as a model for the gaseous disk of a galaxy.

8.1 Generation of Spiral Structures
in Differentially Rotating Shallow Water

Experiments carried out with the configurations shown in Fig. 6.5 yield the following results which we summarize here [8.1–7].

1. Differential rotation of shallow water with a velocity jump between a fast-rotating core and a relatively slow-rotating periphery is unstable both for low Mach numbers $M = (\Omega_1 - \Omega_2)\varrho_0(gH_\eta)^{-1/2} < 1$ and for high Mach numbers $1 \leq M \leq 12$ (here Ω_1 and Ω_2 are the angular velocities of the core and the periphery, respectively).

2. The developing instability gives rise to spiral waves of surface density, crossing the discontinuity (the generator region) and stretching across the core and the periphery. In stationary regimes, the waves form a symmetrical pattern of m arms – elevations on liquid surface (mode m). The pattern rotates steadily in the same direction as the core, at an angular velocity Ω_p (Figs. 8.1, 2). The only exception is the degenerate mode $m = 0$, existing in a relaxatory regime: as it develops, the liquid withdraws from the velocity shear region and the insta-bility ceases; after the wave reflected by the outer wall of the vessel returns, the instability is regenerated again, and so on *ad infinitum*.

3. The spirals are trailing, their outer ends pointing backwards with respect to the direction of pattern rotation.

4. In the experimental regimes with a fast-rotating periphery (Fig. 8.2), the visible length of the spiral arms along their ridges is at least an order of magnitude greater than the Rossby-Obukhov radius $r_R = (gH_\eta)^{1/2}/2\Omega_2$, the counterpart of the characteristic epicyclic radius in galaxies (on the periphery, $\cos \alpha \approx 1$). The arms show no tendency to decay into smaller-scale structures (recall that in real galaxies of the type we are trying to simulate, arm lengths are also much greater

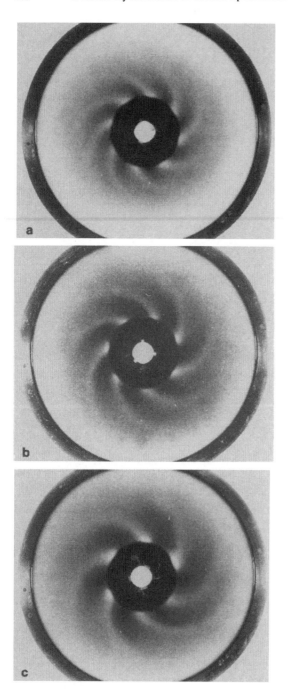

Fig. 8.1 a-c. Spiral waves of surface density excited in differentially rotating shallow water: modes $m = 8$–0 with $\Omega_2 = 0$ (experimental configuration as in Fig. 6.5a; the core and the spiral pattern are rotating clockwise). From *top* to *bottom* **a, b, c**

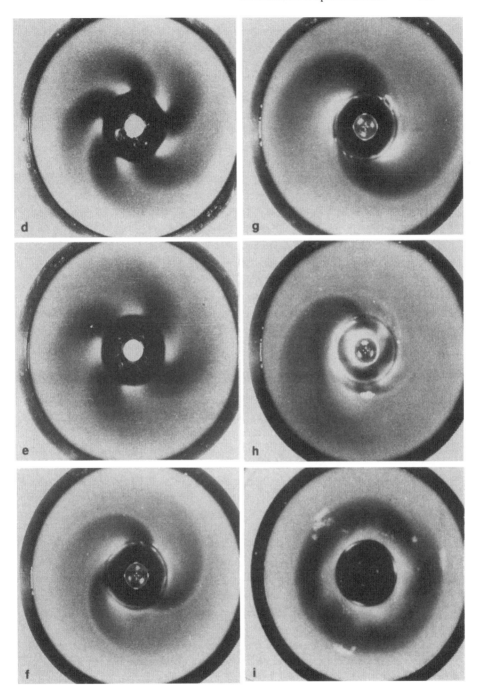

Fig. 8.1 d-i. Spiral waves of surface density excited in differentially rotating shallow water: modes $m = 8$–0 with $\Omega_2 = 0$ (experimental configuration as in Fig. 6.5a; the core and the spiral pattern are rotating clockwise). From *top* to *bottom* on the *left* **d, e, f** and on the *right* **g, h , i**

than $r_R = c_s/2\Omega$ where Ω is the characteristic angular velocity of the gaseous disk in the region occupied by the spiral pattern).

5. The spiral pattern rotates at a rate intermediate between those of the core and the periphery.

6. For a given mode, Ω_p is a monotonously increasing function of Ω_1 (that is, of the Mach number for fixed H_η). As can be seen from Fig. 8.3, this dependence is nearly linear if $m \gg 1$ (Fig. 8.3 corresponds to the case of $\Omega_2 = 0$ but all the features characteristic for this case are also observed when the periphery is rotating).

7. Each of the modes can exist only in a specific range of parameters. For instance, under the conditions of Fig. 8.3, modes $m = 6$, 5, 4, 3 can be realized only if the Mach number does not exceed M = 3.3, 3.7, 4.0, 4.4, respectively. As the system passes these boundaries with gradually increasing M (that is, Ω_1), the current mode is restructured into a larger-scaled one, with less spiral arms on the system perimeter. This mode restructuring is a highly nonlinear process, as indicated, for example, by its abrupt character and the existence of hysteresis. The data in Fig. 8.3 illustrate the general rule that the number of spirals on the system perimeter decreases with increasing M.

8.2 Spiral-Vortex Structures

The experiments indicate that there are vortices between the spiral arms in the shear region, drifting steadily with the arms about vessel axis, so that together they form a joint spiral-vortex structure (Fig. 8.4). Particles of liquid entrapped in the vortices follow closed banana-shaped trajectories whose longitudinal axes are aligned with the velocity jump line. This motion is anticyclonic, against the rotation of the structure as a whole. The maximum radial velocity of particles in the vortices v_ϱ is close to $(gH_\eta)^{1/2}$; the maximum azimuthal velocity of such particles, in the frame rotating with the spiral pattern, is somewhat greater. Liquid layer thickness is the smallest at vortex centers. With a steady anticyclonic motion of the liquid at the ends of the banana, the Coriolis force and the force of hydrostatic pressure are directed towards the vortex center, so the resulting centripetal acceleration is greater than the Coriolis acceleration. In other words, the angular velocity of proper rotation in these areas of the vortices is higher than the angular velocity of the spiral pattern as a whole. Therefore the regime of these vortices is not the Rossby regime given by (2.3) – in contrast to that of the Rossby vortices discussed in the previous chapter. The latter vortices have a slow proper rotation (compared to overall system rotation) and the Coriolis acceleration in them is greater than the centripetal acceleration; this is why Rossby anticyclones, unlike the vortices discussed here, are elevations on liquid surface.

The fact that at the ends of the banana vortices the centripetal acceleration v_ϱ^2/a (where a is the banana half-width) exceeds the Coriolis acceleration $2v_\varrho\Omega_p$ implies that $a < r_R$ where $r_R = (gH_\eta)^{1/2}/2\Omega_p$. This is yet another distinction

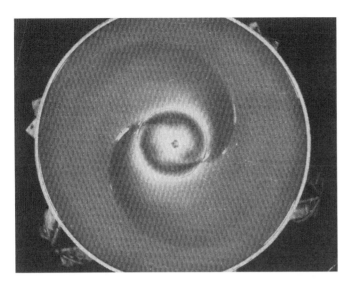

Fig. 8.2. Spiral waves of surface density excited in differentially rotating shallow water: mode $m = 2$ with $\Omega_1 = 13$ rad/s, $\Omega_2 = 2.6$ rad/s, and $H_\eta = 0.35$ cm (experimental configuration as in Fig. 6.5b; the core, the periphery, and the spiral pattern are rotating clockwise; $\Omega_p = 6$ rad/s)

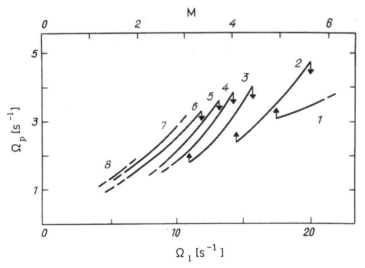

Fig. 8.3. Spiral pattern angular velocity Ω_p as a function of Mach number M, that is, of core angular velocity Ω_1, with $\Omega_2 = 0$ and $H_\eta = 0.2$ cm (experimental configuration as in Fig. 6.5a; mode numbers are indicated for each curve; the limits of existence of a given mode m are marked by *arrows* which indicate the direction of mode restructuring when going over these limits)

of the "interarm" vortices from the large-scale Rossby anticyclones described in the previous chapter.

Unlike the interarm vortices, the spiral arms proper are waves passing through the medium and capturing no particles in a steady rotation with the spiral pattern. Under the influence of these nonlinear waves, particles perform oscillatory motions in the radial direction and shift slowly (as compared to the rotation of the spiral pattern) in the azimuthal direction relative to the bottom (Fig. 8.5). On the periphery, the waves run ahead of the medium ($\Omega_p > \Omega_2$), pushing the transit particles; in the core, the situation is opposite: the waves lag behind the medium ($\Omega_p < \Omega_1$), slowing down particle circumvolution. This gives rise to a very substantial broadening of the initial flow velocity jump which has been the primary cause of the instability (Fig. 8.6).

Fig. 8.4. A spiral-vortex structure excited in differentially rotating shallow water: mode $m = 2$ (experimental configuration as in Fig. 6.5b; the camera rotates with the pattern, that is, $\Omega_c = \Omega_p$)

The character of matter motion relative to the spiral pattern also affects the wave profiles. In the areas where matter enters the wave, a sharp "shock-like" front is observed: it goes along the outer edges of the spiral arms on the periphery and along their inner edges (relative to the center) in the core. In the shear region, that is, at the boundary between the core and the periphery, the azimuthal velocity of particles (both transit and trapped) coincides with that of the wave – that is, a "co-rotation" of the matter and the spiral pattern occurs. Here the particles enter the spiral arm by moving in the radial direction – that is, inwards at the outer edge of the arm and outwards at its inner edge (Fig. 8.4).

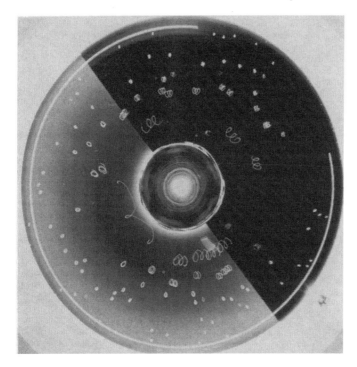

Fig. 8.5. Particle trajectories in the structure shown in Fig. 8.2 (the camera rotates with the periphery, that is, $\Omega_c = \Omega_2$)

To conclude this section, it might be interesting to note that natural spiral structures are observed not only in galaxies but also in hurricanes (typhoons) [8.8–11], see Figs. 2.11 and 8.7. However, while one can speak with a certainty about the wave origin of such structures in the case of galaxies, a further investigation is needed in the case of hurricane spiral arms in order to find out to what extent they are waves and to what extent particle trajectories (see [8.10, 11], particularly [8.11] where a wave mechanism of hurricane spiral generation is discussed, related to the instability of the rotation profile).

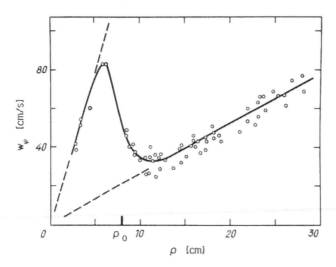

Fig. 8.6. Linear azimuthal velocity w_φ as a function of radius ϱ under the experimental conditions of Fig. 8.2: *circles* are the local experimental values; *solid curve* is the result of averaging over the azimuth; *dashed curves* correspond to rigid body rotation at angular rates Ω_1 and Ω_2

Fig. 8.7. Example of a spiral structure in a typhoon. From [8.8]

8.3 The Common Mechanism Generating Solitary Rossby Vortices in Planetary Atmospheres and Spirals in Galaxies

The whole bulk of experimental evidence discussed above, that is: the very existence of instability in a differentially rotating system, the formation of spiral structures, their trailing shapes, the rotation of the spiral pattern in the same direction as the core at a velocity intermediate between those of the core and the periphery, the qualitative insensitivity of the instability pattern to the peripheral angular velocity, the observed switchover sequence between modes with different numbers of spiral arms as the value of the velocity jump in the flow system is changed, the existence of instability at all values of the Mach number (including $M \gg 2\sqrt{2}$) – all this is in a good agreement with the linear theory where a hydrodynamic mechanism is used to describe the generation of spiral structures in galaxies whose rotation curves contain velocity jumps of the type shown in Fig. 2.13 [8.12–15].

There is also a qualitative agreement between the principal results of our experiments described here and the results of numerical simulations, initiated by these experiments. Such calculations were carried out in [8.16, 17] for the conditions of galaxies and in [8.18, 19] for those of our shallow water experiments. The results of these calculations are illustrated in Figs. 8.8, 9. There is no doubt of the qualitative similarity between these pictures and Figs. 8.1, 2.

The instability responsible for the spiral-vortex structures described here can be identified as the centrifugal instability of a differentially rotating liquid where the inner part is rotating faster than the outer one. As opposed to the K-H instability, existing only for $M \leq 2\sqrt{2}$ (Sect. 4.1), this kind of instability is also observed for much greater values of M – even for $M \approx 12$. (As before, we speak here of the regimes where steadily rotating spirals are formed. For $M > 12$, the steady pattern breaks down spontaneously and is replaced by relaxatory regimes with $m = 0$. The maximum value of M implemented in the experiments is $M \approx 20$).

The centrifugal instability of differentially rotating shallow water for high Mach numbers is, as far as we know, a new and hitherto unknown type of instability, discovered in the series of theoretical and experimental works described here [8.1–7, 12–16].

Thus, the centrifugal instability can create both the spiral-vortex structures described here – under certain experimental conditions – and the Rossby vortex chains described in the previous chapter (solitary vortices, in particular) – under certain other conditions. The laws governing the competition and switchover of instability modes with different numbers of spiral arms (Fig. 8.3) and with different numbers of Rossby vortices (Fig. 7.9) are similar – see the auxiliary Fig. 8.10. The bulk of the above evidence allows one to make the conclusion that the generation of such different natural structures as the largest solitary

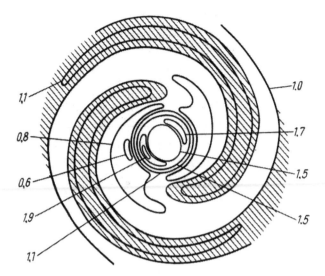

Fig. 8.8. Nonlinear stage in the development of hydrodynamic instability mode $m = 2$ of a differentially rotating gaseous disk: isolines of perturbed surface density for the spiral structure into which the initial perturbation evolves after 4.3 core revolutions; the *numbers* at the curves are the ratios of perturbed surface density to its unperturbed value (numerical calculation for $\delta/\varrho_0 = 0.1$; $\Omega_2/\Omega_1 = 0.1$; M=4.34). From [8.16]

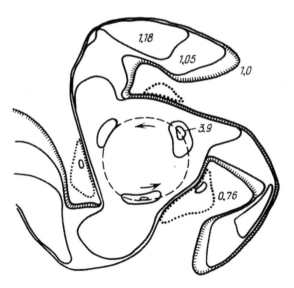

Fig. 8.9. Nonlinear stage in the development of hydrodynamic instability of differentially rotating shallow water: isolines of perturbed layer thickness after four core revolutions; the *numbers* at the curves are the values of relative thickness h (numerical calculation for $\delta/\varrho_0 = 0.1$; $\Omega_2 = 0$; M=3). From [8.19]

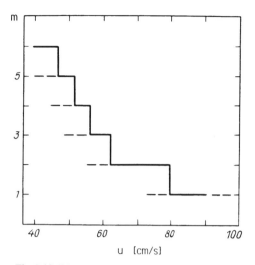

Fig. 8.10. Right-hand limits for the existence of spiral patterns with different number of arms m with changing (increasing) flow velocity shear between the core and the periphery under the experimental conditions of Fig. 8.3

atmospheric vortices and the spiral structures in galaxies is governed by the same physical mechanism, the centrifugal instability of flows with anticyclonic velocity shear.

Evidently, what has just been said does not mean that the centrifugal instability is the sole cause to produce large atmospheric vortices. Cyclones (for example, such as Jupiter's Barges) are generated by unstable sheared flows where the outer region is moving faster than the inner one. In such a situation, a non-centrifugal instability is the working agent; we do not correlate it with the generation of spiral structures in galaxies.

8.4 Asymmetrical and Outbranching Spirals

All the spiral structures which have been demonstrated so far are observed with the core and the periphery rotating steadily. Let us now consider structures which are observed in transient (nonstationary) regimes – during mode switchovers $m_1 \rightarrow m_2$, occurring when core angular velocity is rising smoothly. For a given value of m_1, the switchover scenario depends on how Ω_1 changes with time. During the switchover, the spiral-vortex structure is self-oscillating: the dimensions, shapes, intensities, and angular velocities of individual vortices and arms are changing periodically with time. The changes are characterized by ever growing amplitudes and by phase shifts between the individual elements of the structure. Adjacent vortices move to and from each other, join and part, until they finally merge to give rise to the new stationary mode m_2. The duration of mode switchover may be up to several dozen core revolutions, depending on m_1 and m_2.

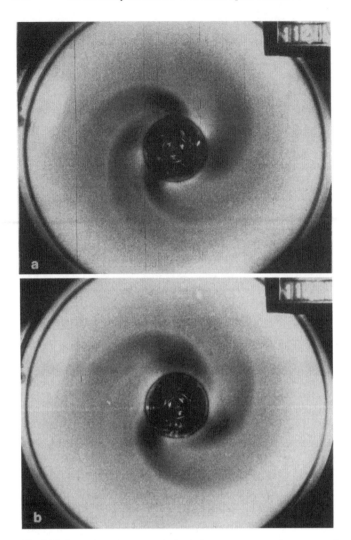

Fig. 8.11. Examples of asymmetrical spiral structures appearing at a certain stage of transition from mode $m = 4$ to a larger-scale mode as the core rotation velocity increases under the experimental conditions of Fig. 8.1

Within this general picture, two important circumstances should be emphasized. First, during the entire switchover process, the spiral arms are different in amplitude, angular velocity, and pitch angle, thus the transient spiral pattern is usually asymmetrical (Fig. 8.11). Second, arms that are twisted more tightly have higher angular velocities; for this reason, at certain stages of the switchover the spiral pattern shows outbranching arms, when one arm overtaking another has already merged with it in the velocity jump region but still remains separated from it on the periphery (Fig. 8.12). One may suggest on this ground that the galaxies with asymmetrical spiral patterns and/or outbranching arms have their cores rotating nonsteadily and are in the process of mode switchover. This hypothesis is in a good agreement with the modern views on the age of spiral galaxies: they are supposed to be no older than 100 to 200 central core revolutions, which, judging by the results of our simulation experiments, is just a few times the duration of mode switchover. This may be the explanation for the observed abundance of galaxies displaying asymmetrical and outbranching arms (Sect. 2.4).

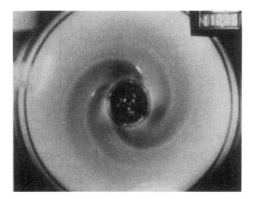

Fig. 8.12. Example of a spiral structure with outbranching arms, appearing at a certain stage of transition from mode $m = 4$ to mode $m = 2$ as the core rotation velocity increases slowly under the experimental conditions of Fig. 8.1

8.5 The Influence of Viscosity and Friction

It is sometimes argued that our experimental configurations are greatly different from the natural objects to be simulated because they have a solid bottom. The friction of the working liquid against vessel bottom is certainly a new effect, not encountered in real galaxies. Besides, the viscous (Eckman) bottom layer, where the velocity is sheared along the vertical, may be unstable and thus may act as a generator of large-scale structures. So one might ask if the spiral-vortex structures observed in our experiments could be produced by the instability of the Eckman layer. One might also ask to what extent the laboratory situation where the generated structures are subject to friction against the bottom is adequate to the actual situation in spiral galaxies which have no "bottom".

Our answers to these questions are as follows. First, the results described above remain practically the same over the entire range of working liquid viscosities. A change in viscosity does somewhat displace the boundaries of the regimes in which the respective azimuthal modes exist. This change is associated with hysteresis – just like a change in any other system parameter. One may conclude that although some viscosity is necessary in the above experiments (for transmitting the rotation of vessel bottom to the liquid layer), its magnitude has little effect on the outcome of the experiments with differentially rotating shallow water. This is supported by the following estimate: the ratio of the viscous term in the Navier – Stokes equation to the term associated with the gradient of hydrostatic pressure is $\nu|\nabla^2 v|/|\nabla p| \approx (\nu v_0/H_\eta^2)(gH_\eta/l)^{-1} \approx 0.1$ where $\nu = 0.05 \text{ cm}^2/\text{s}$, $v_0 \approx (gH_\eta)^{1/2}$ is the characteristic velocity of the liquid with respect to the bottom, $H_\eta \approx 0.3$ cm is a typical value of liquid thickness in the experiments, and $l = 2$ to 3 cm is a typical transverse dimension of a spiral arm. Consequently, to a first approximation one can neglect the viscosity of the medium since it does not play any decisive role in the development of the instability which produces spiral structures. Accordingly, all the effects described in this section remain qualitatively the same when viscosity is increased by about an order of magnitude, although the typical value of the Eckman number is not very small (E ≈ 0.1). Only when the Eckman number approaches unity do the spiral-vortex structures cease to be produced. One may thus conclude that a value of E ≈ 0.1 can be considered sufficiently small for our present purpose. (In the experiments with Rossby vortices (Sect. 9.5), the maximum Eckman number is about $2 \cdot 10^{-2}$ which is certainly a small enough value as the results show.)

Second, the generation pattern of the spiral-vortex structures described above is entirely distinct from what is observed for spirals of quite different origin – that is, those produced by the instability of the Eckman layer with a vertical shear of rotation velocity [8.20, 21]. The latter spirals are twisted very tightly and are produced under quite different experimental conditions. For an illustration, compare Figs. 8.1, 2 with Fig. 8.13 showing an example of such "viscosity" spirals [8.21] (we do not dwell here on the mechanism of their generation).

Third, as indicated above, the main qualitative results of the simulation experiments in question do not depend on periphery rotation velocity and, in particular, on whether the periphery is rotating or not. In the latter case, however, there is no Eckman layer at all and the vertical profile of liquid velocity is certainly not what it is in the case of $\Omega_2 \neq 0$.

Fourth, a special control experiment was performed in order to check the two-dimensionality of the above spiral-vortex structures. In this experiment, the shallow water is simulated with a thin layer ($H_\eta \approx 0.3$ cm) of a light oil with a viscosity close to that of the $NiSO_4$ working solution commonly used in the experiments. A thick layer (≥ 1 cm) of a heavier liquid not mixing with the oil is placed between the oil layer and vessel bottom (the configuration shown in Fig. 6.5b). The "substrate" liquid (glycerin) is so viscous that, drawn into a differential rotation by the bottom, its flow is always stable over the entire working range of Ω_1 and Ω_2. In fact, the presence of the substrate contributes a single new feature to this control experiment: the initial width of the velocity shear in the working (upper) liquid layer now depends more on substrate thickness than on the value of H_η. The experiment shows that a thin layer of differen-

Fig. 8.13. Spiral structures (vortices) produced as a result of boundary layer instability developing over a disk (40 cm in diameter) rotating counter-clockwise in the air at 1800 rps. The structures are visualized by ejecting some lightly colored gas from the black surface of the disk. The axes of the spiral vortices lie in the plane *perpendicular* to the system rotation axis. From [8.21]

tially rotating liquid with a free surface is unstable under these conditions, too, producing spiral-vortex structures not very different from those described above. The motion of the working liquid in the structures is visualized with two types of test particles, the lighter particles floating upon the free surface as before and the heavier ones near the substrate in the working liquid. A direct visual observation shows that the vortices between the spiral arms at the lower boundary of the working liquid are practically the vertical projections of those existing upon the free surface. So this control experiment may be considered a good test for the two-dimensionality of the structures under investigation.

Fifth, the switchover sequence of azimuthal instability modes described in the previous sections of this chapter (the number of arms in the structure becoming less as the velocity of the sheared flow is increased) is a manifestation of a general law which is also observed in our experiments on the generation of anticyclonic Rossby solitons, where the thickness of the working liquid has been up to 40–50 Eckman layers (Sects. 7.1–3). The same rule is also observed in the experiments on producing Rossby cyclones [8.22, 23], where the layer thickness (20 cm) is about 10^3 Eckman layers and the working liquid itself is incompressible not only in the three-dimensional but also in the two-dimensional sense (it has no free surface). A similar result has been obtained in the experiments reported in [8.24, 25] despite their absolutely different conditions (see the end of Chap. 7). All this implies that this general rule can be in no way associated with the instability of the Eckman layer.

Finally, sixth, as shown in Sect. 4.1, our simulation experiments with shallow water resting upon the vessel bottom yield results which are physically absolutely identical to those obtained by Landau for a two-dimensional gasdynamic medium. We see that in this case, too, the presence of a bottom does not significantly affect the outcome of the experiments. A similar result was obtained in the experiments cited above [8.24, 25] where the dynamics of the K-H instability developing in a gaseous medium rotated differentially by horizontal lids is adequately described by a two-dimensional theory.

Thus, the observed sequence of instability mode switchovers in flows with velocity shear is qualitatively the same in experiments carried out with very different values of the Rossby, Eckman, Froude, and Mach numbers (ranging from Ro $\ll$ 1 [8.22, 26] to Ro $>$ 1 [8.2], from E $\approx$ 10^{-6} [8.22] to E $>$ 1 [8.24], from F $\rightarrow$ 0 [8.22, 24] to F $\gg$ 1 [8.27], and from M $\rightarrow$ 0 [8.22] to M $\geq$ 10 [8.28]). This sequence, illustrated in Fig. 7.9, corresponds, as shown in Sect. 7.3, to a simple criterion: a maximum of the linear instability increment, assuming an approximately constant decrement of viscous damping. This concept corresponds well to systems with an "external" friction – that is, a Rayleigh (or, more precisely, Eckman) friction against an underlying surface, acting as the primary channel of structure dissipation in shallow water, see (7.4–8). For example, the external (Rayleigh, Eckman) friction in shallow water with layer thickness H is much stronger than the inner viscosity which is inversely proportional to the square of the wavelength λ ($\lambda \gg H$). The terrestrial atmosphere, in particular, is a system with predominating external friction against the underlying surface – with respect to planetary-scale vortices [8.29, 30].

But there seems to be no external friction in galaxies. Our hypothesis is that *external friction does exist in galaxies*. It is the friction between the gaseous structures and the stars. [Recall that, according to the hydrodynamic concept we discuss here, the spiral structures are formed in the *gaseous* subsystem of the galactic disk.] If, for instance, all the stars were located in the middle plane of the galaxy, the friction of the spiral gaseous structures against the stars would be an obvious analog of the Eckman friction between atmospheric (say, terrestrial) vortices and the underlying surface. Clearly, the fact that stars are actually distributed over galactic disk height does not affect the qualitative picture. As regards the quantitative estimates, a special analysis is needed. In particular, in order to extend the external friction theory to the case we are discussing, one has to show that the specific volume force of gas friction against the stars in a galactic disk can be represented in the same Rayleigh form as the friction of the atmosphere against a terrestrial surface, namely, $F = \varrho v / \tau$ where ϱ is the density of matter, v is the velocity at which the gas of the structures is moving with respect to the stars, and τ is the characteristic time of velocity dissipation. The assumption about a Rayleigh-type friction of gaseous galactic disks appears quite natural. It is useful to point out in this connection that the same friction law, as we shall see later, is also assumed when analyzing the dissipation of large-scale zonal flows in upper Jovian atmosphere, see Supplement S10.1. The underlying surface is then assumed to be the surface of the thick adiabatic gaseous layer located below the upper layer of clouds which bears large-scale vortices like the Jovian Great Red Spot.

Thus the friction of rotating shallow water against the bottom of the vessel used for laboratory simulation of the mechanism which generates spiral structures in galaxies is not an interference but, on the contrary, a necessary prerequisite for such a simulation to be adequate.

To conclude this section, we note that the spiral structures considered in this book also have nothing to do with the so-called Görtler-Taylor instability related to liquid motion over a concave surface (the common view holds it that this kind of instability is responsible for the generation of spiral waves in baths) [8.31–33]. The absence of any connection between the Görtler-Taylor instability and the spiral structures demonstrated in this book follows, for example, from the fact that they can be perfectly well generated even when the bottom of the experimental vessel has no concave areas – for instance, if the first configuration option is used (Fig. 6.5a) where the core and the periphery are cones rather than paraboloids and the transition surface between them is convex.

8.6 Laboratory Simulation and Astronomical Observations. Predictions for Astronomers

First of all, it must be pointed out that spiral structures with different numbers of arms are found in real galaxies, similarly to the laboratory observations described here. In particular, the natural structures corresponding to mode $m = 0$ are known as ring-galaxies [8.34].

The next principal point is to compare (a) the distribution of shallow water surface density in spiral structures of the type shown in Fig. 8.2 with the distribution of matter density in a real spiral galaxy; (b) the shapes of the spiral arms observed in the laboratory and in the nature; and (c) the relative positions of the co-rotation circles where the angular velocity of matter is the same as that of the spiral pattern. For this comparison we shall use the data on a particular galaxy given in [8.35]. The author of that study which was stimulated by our experiments has processed the observations of ionized hydrogen distribution in the isolated galaxy NGC 1566 which has no satellites, possesses an N-shaped rotation curve and a clearly outlined symmetrical spiral structure of the "grand design" type. The results are presented in Fig. 8.14 where the distribution of zones with different HII brightness is superimposed over the spiral pattern plotted on the basis of the surface brightness profile of the visible spiral structure. The brightest HII zones make up the front lines of star creation, which (similar to the "shock" fronts in the spiral waves observed in the shallow water experiments) stretch along the arms, going over from outer to inner sides of the arms on their way to galaxy center. These transition areas are identified in [8.35] as the areas of co-rotation (they are located at the radius where, according to the more recent results reported in [8.36], the velocity of galaxy rotation drops abruptly). The comparison of Figs. 8.14 and 8.2 shows a high degree of geometrical similarity between the galactic and experimental structures; the position of the co-rotation circle with respect to these structures is virtually the same, too (see also Fig. 8.15).

The comparison further shows that the distribution of shallow water surface density corresponds well to the distribution of ionized hydrogen density over the spiral pattern of the galaxy. In particular, there are areas of very low gas density between the arms of NGC 1566, in the region of the co-rotation circle, which clearly correspond to the location of anticyclonic vortices with low surface density found in the model experiments. This fact is indirect evidence that the predictions about the existence of anticyclones in the gaseous disks of spiral galaxies, made on the basis of our experiments, seem to be confirmed by astronomical observations.

A quite recent study [8.37] may be considered the first direct evidence for the existence of the galactic vortices we are discussing. The study was specially intended to verify the hydrodynamic concept of the genesis of galactic spiral structures. Using the six-meter reflector at the USSR Special Astrophysical Observatory, the field of velocities and the distribution of HII were investigated for the dual-arm galaxy NGC 931 which has a jump on its rotation curve. The authors of [8.37] believe that the areas predicted on the basis of laboratory simulation – that is, areas of anticyclonic matter motion and low HII density – have been identified quite unambiguously between the spiral arms.

From the astrophysical standpoint, it is exactly this kind of predictions that is the most valuable result of the simulations since they help astronomers to formulate a consistent program for observations. The material presented in this chapter allows us to make the following predictions for stationary spiral galaxies exhibiting a velocity jump between its core and periphery:

1) There must be a correspondence, such as illustrated by Figs. 8.14 and 8.2, between the shapes of the spirals observed in galaxies and those obtained in the simulation experiments.

2) A similar correspondence must also exist between the respective surface density distributions of the matter in the spiral arms and in the interarm regions. As we have seen above, the correspondences indicated under (1) and (2) appear to actually exist.

3) The number m of spiral arms in real galaxies must be directly related to the Mach number associated with the jump of rotation velocity: the higher M, the smaller m.

4) Intensive anticyclonic vortices must be observed between the spiral arms in the region of the velocity jump. This is one of the crucial points. As a matter of fact, the gravitational concept also allows the existence of vortices (that is, areas where particle trajectories are closed in the frame rotating with the pattern) in a system of spiral waves. However, the latter concept, within the framework of the single-particle approximation, predicts such vortices (for the same wave amplitude, corresponding to the observations) to be located at the spiral arms, not in the interarm regions [8.38]. The divergence between the gravitational and hydrodynamic concepts in this point is due to the fact that the "potential wells" related to perturbations of the gravitational potential are located *at the arms* where matter density is the highest while the potential wells related to perturbations of the hydrostatic pressure are located *between the arms* where that density is the lowest. Therefore, direct astronomical observations of vortex localization can give a straight answer to the question which of the two concepts is adequate.

5) Co-rotation must exist at that location in the spiral arm where the radial velocity of the matter (participating in the vortical motion) reaches its maximum.

6) Asymmetrical and outbranching spirals must be observed in nonstationary galaxies.

Discussing the hydrodynamic concept, one might ask how often galaxies with a velocity jump between their core and periphery (Fig. 2.13) are encountered among spiral galaxies. This question has two different aspects, that of simulation (numerical calculations, laboratory experiments) and that of observations. According to the numerical estimates [8.16], an initial velocity jump even as low as several percent is already sufficient for a spiral structure to develop on an astronomical time scale. As the structure develops, its velocity jump is smoothed out, as indicated by the results of our laboratory experiments (Fig. 8.6). As regards the observational aspect, the astronomers tended to believe prior to our experiments that rotation velocity jumps of the type shown in Fig. 2.13 are relatively rare, to be encountered in no more than 10–15% of spiral galaxies. The laboratory experiments discussed here have stimulated a special investigation of this problem as carried out at the Special Astrophysical Observatory with the world's largest reflector [8.39–42]. The analysis revealed that many of the previous observations had been made with certainly insufficient spatial resolution and were not accurate enough in velocity measurements. After a new program using improved techniques was carried out, quite different results were obtained: it turned out that velocity jumps of the type shown in Fig. 2.13, sufficient for the formation of spiral structures, are found in a clear majority of spiral galaxies (in no less than 60% of the forty galaxies studied in [8.39–42]). This strengthens our belief that the mechanism of spiral structure genesis discussed here and simulated by our experiments pertains to galaxies of the most common type. The stimulation of the very fruitful observational series mentioned above may also be considered an example of the real benefit gained by astrophysics from the hydrodynamic simulation described here.

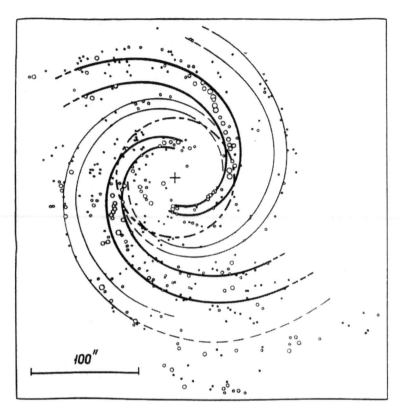

Fig. 8.14. Spiral pattern and distribution of the zones of ionized hydrogen HII in the galaxy NGC 1566: *solid curves* approximate the visible shape of spiral arms as determined from the surface luminosity profile of the galaxy; locations of the HII zones are indicated by *circles* (consecutively larger circles correspond to radiation flux increasing in a geometrical progression); *dots and dashes* outline the corotation circle. From [8.35]

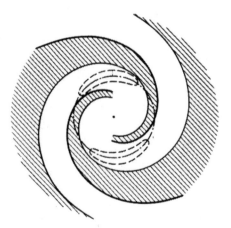

Fig. 8.15. Spiral arms (*hatched*), corotation circle (*dots and dashes*), and the supposed locations of anticyclonic vortices (*dashed curves*) in the galaxy NGC 1566 (plotted in the scale of Fig. 8.14 to facilitate comparison with Figs. 8.2, 4)

We have discussed in this section a particular – centrifugal – kind of hydrodynamic instability as the cause underlying the generation of galactic spiral structures. Another type, the wind instability, is proposed in [8.43, 44] for the interpretation of the mechanism of spiral structure genesis. The comparison of that theory with the experimental data described here calls for a special study.

Worth mentioning is also the investigation published in [8.45] where dual-arm galactic spirals are interpreted through a mechanism related to the motions of individual particles rather than to collective interactions.

Additional explanations for some of the material discussed in this chapter are given in Supplement 8.1.

9. Rossby Vortices and Solitons in Free Motion

The theory of anticyclonic vortical Rossby solitons has been discussed in Chap. 5. The solitonic concept of anticyclonic Rossby vortices has been used as a foundation for the theoretical solitonic models of the JGRS and similar long-lived vortices in the atmospheres of giant planets and in the oceans. The first consistent solitonic theory of the JGRS was proposed in 1976 (see below). In the light of these theoretical results, it appeared quite important to obtain a Rossby soliton in laboratory experiment. The present chapter describes how this problem was solved.

9.1 A Short History of the Experiments

A Rossby soliton was produced for the first time and thoroughly studied in the experiments reported in [9.1–12]. The problem formulation preceding these experiments and the necessary criteria an experimental configuration should comply with have been described in Chap. 6. Here we shall describe the experiments where Rossby solitons are produced and observed in the "free-travel" regime, i.e., with no pumping from zonal flows.

Chronologically, this kind of experiments – with Rossby solitons in the free-travel regime – was the first stage in our hydrodynamic simulation of natural astrophysical phenomena. In those experiments, shallow water was rotated as a body together with an approximately paraboloidal vessel. Monopolar Rossby solitons were obtained, essentially of the type predicted by the theory based on the scalar (KdV) nonlinearity [9.13–18]. In agreement with the theory, the solitons were found to be anticyclones of an approximately round shape, to have dimensions exceeding the radius r_R, to drift with respect to the vessel counter to its rotation at about the Rossby velocity, and to decay gradually in about the viscous damping time (which was close to 10 s in the first experiments with the small paraboloid).

On the other hand, those experiments also yielded new results, largely unexpected from the standpoint of the then existing "purely wave" Rossby soliton theory. In particular, we discovered several features of the Rossby solitons such as their clearly vortical nature (they contain trapped particles of the liquid and carry them along), an unusual combination of size and amplitude (mostly in strong disagreement with the analytical solution given by (5.29,30)), the merging of solitons upon collision, and their stability in spite of the variation of the

Rossby velocity along the meridional coordinate. These findings stimulated development and re-evaluation of the theory and eventually led to a new theory [9.19–25], relying to a large extent on numerical calculation. The new theory is based from the very beginning on the fact, discovered in the experiments, that an area of closed streamlines, containing trapped particles, exists in a Rossby soliton. The results obtained within the framework of this theory are in a good agreement with the whole bulk of experimental evidence.

The second stage of the experiments was concerned with coherent vortical structures in a new system geometry – with *differentially* rotating liquid, where counterflows physically similar to the zonal flows in planetary atmospheres were superimposed on the overall system rotation. At this stage of the experiments, already described in the preceding chapters, two kinds of instability of sheared flows were discovered: the K-H instability and the centrifugal instability. They give rise to chains of large-scale vortices (larger than the Rossby-Obukhov radius), stretching along the system perimeter; under certain conditions, these vortices are Rossby solitons.

Later, in the third stage of the experiments, described in Chap. 7 and carried out with the modified configuration, we found such a regime of zonal flows in which their instability generates a new nonlinear vortical structure, the Rossby autosoliton. The latter is a solitary, undamped, large-scale ($a > r_R$), anticyclonic vortex, unique over the entire system perimeter, self-organizing in the flows; it is a stationary physical solitonic model of natural vortices such as the JGRS. Thus the solitonic concept of the JGRS and other physically similar vortices was validated experimentally. These experiments provided a simple explanation of the fact that all of the large vortices on Jupiter and Saturn are anticyclones, with a few exceptions such as the Brown Ovals of Jupiter and the UV Spot of Saturn, and also made it possible to explain the physical nature of these exceptions. This gave rise to the hypothesis that the large-scale cyclonic vortices which exist on Jupiter and Saturn as rare exceptions might have the same origin and the same mechanism of generation as the cyclones which are typical for Earth.

Along with the experiments on producing Rossby solitons by unstable zonal flows, other experiments (also with differentially rotating shallow water) were carried out, aimed at simulating the hydrodynamic mechanism of spiral structure generation in galaxies with a velocity jump in their rotation profile. As a result, we came to the conclusion that the both natural phenomena – the large-scale vortices in the atmospheres of giant planets and the spiral structures in galaxies of the above type – are probably generated by the same physical mechanism based on the centrifugal hydrodynamic instability of differentially rotating shallow water where the core is rotating faster than the periphery. This stage of the experiments has been discussed in the previous chapter.

Identification of the solitary anticyclonic Rossby vortices under study as solitons was made in the experiments described in the present chapter.

9.2 Rossby Solitons in the Laboratory and Their Properties

We now turn to the experiments on producing and studying Rossby solitons in the free-travel regime.

The first thing to be taken into account is the expected size of a Rossby soliton. According to the theory presented in Chap. 5, the characteristic soliton radius a must be greater than the Rossby-Obukhov radius r_R. This determines the design of the local generators used for producing Rossby solitons. The main results obtained with such generators (described in Sect. 6.2) and with the diagnostic techniques (described in Sect. 6.5) are given below.

Fig. 9.1. An anticyclonic Rossby soliton in the small paraboloid with $H_0 = 0.5$ cm=const (liquid motion and surface elevation in the vortex are both visualized). The picture was taken 3 seconds after the pumping disk was switched off. The vortex is drifting clockwise, counter to vessel rotation

Figure 9.1 shows a typical anticyclonic vortex generated with the pumping disk about 3 s before the picture was taken, drifting at constant layer thickness against vessel rotation. Its profile (Figs. 9.2, 3) is approximately Gaussian, falling off much more steeply than that of a classical vortex where $v \propto r^{-2}$. The vortex diameter $2a$ (at mid-profile of h) is about $2.5r_R$, its relative amplitude is $h_0 \approx 0.5$, the characteristic angular velocity of proper rotation (at the same mid-profile) is about 1/4 the angular velocity of system rotation. An object with such parameters is a geostrophic vortex in the Rossby regime (2.3), more or less corresponding to the interval of maximum dispersion on the curve in Fig. 5.4a. Vortex drift at H_0 = const is directed westwards, i.e., in the same direction as the drift of Rossby waves. Figure 9.4 shows another anticyclonic vortex, produced also with H_0 = const by the pumping disk in position 1, dyed from above in position 2, drifting clockwise and photographed 18 s after it was created. Its parameters

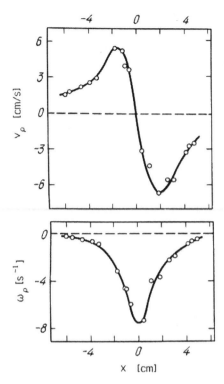

Fig. 9.2. Spatial profiles of proper rotation of a Rossby anticyclone in the latitudinal cross-section of the paraboloid: meridional component of linear velocity v_ϱ and angular rate $\omega_\varrho = v_\varrho/x$ as functions of the coordinate x measured from vortex center along the parallel in the direction of vessel rotation (the experimental conditions are close to those of Fig. 9.1)

are approximately the same as those of the vortex in Fig. 9.1. The vortex drifts for 20 s at an average velocity $V_{dr} \approx 2$ cm/s $\approx 0.9 V_R$ where V_R is the Rossby velocity for $H_0 = $ const given by (6.5). This value of drift velocity is about three times lower than the typical linear velocity of vortex rotation measured at profile maximum (Fig. 9.2). Figure 9.4 illustrates a fundamentally important property of this vortex: the transport of captured liquid particles. This is typical for all such vortices of sufficiently large amplitudes.

Vortex drift velocity increases with liquid layer thickness and with vortex amplitude $(\delta H)_0$, in qualitative agreement with (5.33). The drift velocity increases considerably if the vessel is rotated faster and decreases as it is slowed down (Fig. 9.5). At a certain angular velocity of vessel rotation ($\Omega < \Omega_0$), the vortex is at rest with respect to the vessel, and at still lower velocities it drifts in the opposite direction, eastwards (Fig. 9.6). These features (in particular, the value of the angular velocity shift corresponding to zero drift), observed both with the small and the large paraboloids, are in good agreement with (6.7). Thus one can see that these vortices are actually Rossby vortices. Figure 9.7 illustrates the

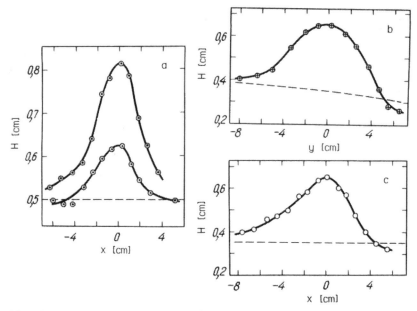

Fig. 9.3. Spatial profiles of layer thickness in an anticyclone excited in the small paraboloid by technique "a": (**a**) azimuthal profiles for different vortex amplitudes with $H_0 = 0.5$ cm = const; (**b, c**) meridional and azimuthal profiles of a vortex with $\Omega = \Omega_0$, $H_0 = 0.35$ cm at $\varrho = 10$ cm $\approx D/2\sqrt{2}$ (*dashed lines* show the level of unperturbed free surface; the azimuthal coordinate x increases in the direction of vessel rotation – counter to vortex drift; the meridional coordinate y increases towards vessel center)

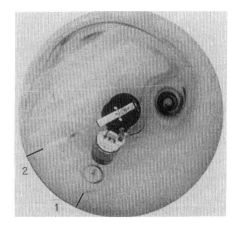

Fig. 9.4. An anticyclonic vortex drifting against vessel rotation (the small paraboloid, $H_0 = 0.5$ cm = const). The vortex is created in clear water by the pumping disk (*1*) and drifts clockwise. Liquid dye is injected into it in position (*2*). The streak of dye, initially oriented along vessel radius, is twisted into a spiral by the proper (differential) rotation of the vortex and moves clockwise along the parallel together with all the rest of liquid captured by the vortex. The picture was taken 18 s after vortex creation; vortex lifetime is about 20 s

dynamics of a Rossby vortex produced by jet injection according to technique "b" and also having the anticyclonic polarity.

In a control experiment, unlike all the experiments described so far, the vessel was made exactly paraboloidal so that at a certain nominal angular velocity all points of the equilibrium liquid surface were at the same distance H_η from the bottom along the rotation axis. The layer thickness H_0 (measured along the normal to liquid surface) then became a function of location, $H_0 = H_\eta \cos \alpha$, and the Rossby velocity was zero as implied by (6.8). The experiment has shown accordingly that the vortices (excited by technique "b") stand still when the vessel is rotated at the nominal velocity, drift westwards when the velocity is higher and eastwards when it is lower, which is in agreement with the theoretical relationship (6.9) [9.12].

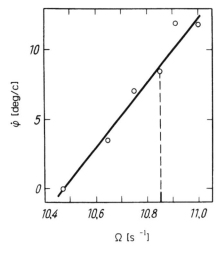

Fig. 9.5. Anticyclone angular displacement rate $\dot{\varphi}$ with respect to the paraboloid (in degrees per revolution of the paraboloid) as a function of the angular rate of the paraboloid Ω for $H_0 = 0.5$ cm at $\varrho = 10$ cm $\approx D/2\sqrt{2}$ (the small paraboloid: $H_0 = $ const at $\Omega = \Omega_0 = 10.85$ rad/s, the gradient of H_0 is directed outwards for $\Omega > \Omega_0$ and inwards for $\Omega < \Omega_0$)

The sum of the properties exhibited by the vortices in question, namely: (a) size $a > r_R$; (b) steep velocity profile; (c) drift in the direction of Rossby wave propagation at a velocity close to V_R and much higher than the drift velocity of a linear Rossby wave packet; (d) free-run path much greater than that of a linear wave packet given by (5.17) – all this taken together makes one suppose that the observed anticyclonic vortices are monopolar Rossby solitons. We shall be able to draw a more definite conclusion somewhat later, after we compare the behavior of cyclones and anticyclones.

The experiments show that the dimensions of the longest-lived vortices among those studied are in the range $r_R < a < 3r_R$. Vortices of smaller size ($a \lesssim$

Fig. 9.6. A Rossby anticyclone drifting along vessel rotation (counter-clockwise) due to an inward gradient of H_0 (the large paraboloid, $H_0 = 1$ cm at $\varrho = 25$ cm $\approx D/2\sqrt{2}$, $\Omega = 7.1$ rad/s; the vortex is created by technique "a")

r_R) decay rapidly. So do vortices whose sizes are too large – larger than the intermediate geostrophic radius r_{IG} (5.32), i.e., with $a \gtrsim 3r_R$ ($r_{IG} \approx 3r_R$ under the conditions of our experiments) [9.2, 3].

The experiments indicate that a more or less arbitrary initial extensive perturbation in the liquid with a high enough amplitude quickly evolves into well-developed Rossby vortices. In particular, this is seen in Fig. 9.8 where the consecutive positions of the same vortex are shown for the indicated time intervals. Thus a large-amplitude Rossby vortex of the type we are discussing is a preferable, or "attractive", solution. To put it otherwise, an anticyclonic Rossby vortex can be said to manifest attractor properties, or the property of self-organization, which is an additional evidence of its stability. We shall show in Chap. 11 that a similar attractor property is also manifested by dipolar Rossby vortices.

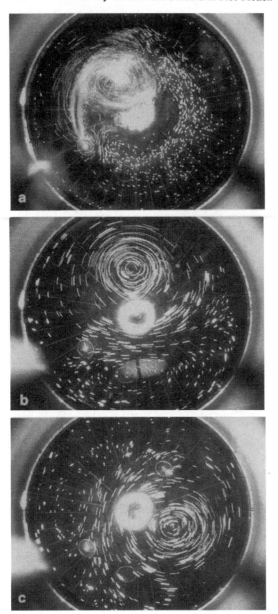

Fig. 9.7 a-c. Drift of an anticyclone produced by technique "b" (the small paraboloid; the time intervals are 10 s between frames **a** and **b**, 5 s between **b** and **c**; in position **b**, the vortex has accomplished a full revolution around the vessel which is rotating counter-clockwise)

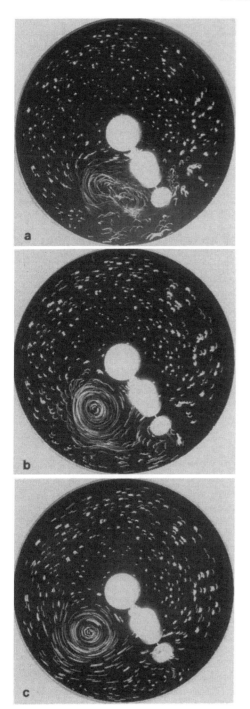

Fig. 9.8 a-c. A Rossby anticyclone developing from an irregular perturbation created with the pumping disk (the small paraboloid, $H_0 = 0.3$ cm; time intervals between the frames are 1.2 s each)

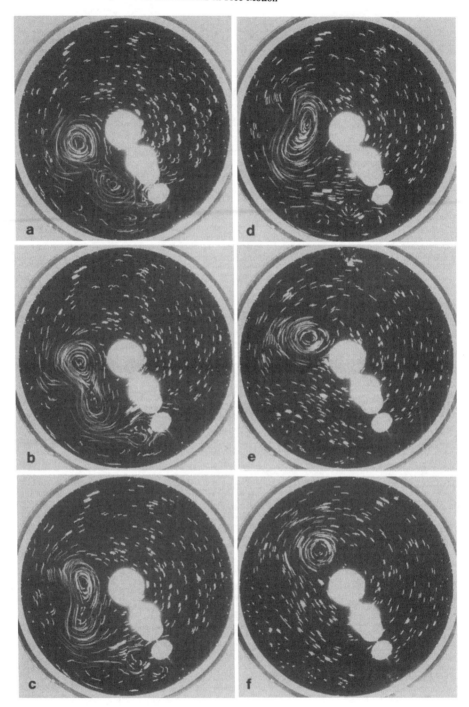

Fig. 9.9 a-f. Convergence and merging stages of two Rossby anticyclones created with the same pumping disk (the small paraboloid, $H_0 = 0.3$ cm; the time intervals are 0.6 s between frames **a**, **b**, **c**, and **d**, 1.8 s between frames **d**, **e**, and **f**)

9.3 Collisions of Rossby Vortices

The experiments reported in [9.5] have shown that large-amplitude Rossby vortices carrying entrapped particles of the liquid collide inelastically: they either merge into a single vortex or jointly dissolve, creating a flow. The first scenario prevails when the vortices converge quickly enough so that the characteristic time of their mutual approach is not large compared to the period of their proper rotation. The second scenario prevails when the vortices converge more slowly. An example of the merging of two Rossby anticyclones following one another is shown in Fig. 9.9: the vortex with the larger amplitude moving at first behind the weaker one is catching up with it.

It must be pointed out that cyclones also tend to merge upon collision; in other words, merging is possible for Rossby vortices of like polarities. This property is also exhibited by long-lived natural vortices in the atmospheres of Jupiter and Saturn [9.26, 27].

Recall that all the vortices we have been studying carry trapped particles with them, hence they have large enough amplitudes. As we shall see later (Sect. 9.6), it is this circumstance that is responsible for the inelasticity of Rossby vortex collisions.

9.4 Cyclone-Anticyclone Asymmetry

All of the Rossby vortices illustrated so far (Figs. 9.1–9) have been anticyclones. An example of a cyclone is shown in Fig. 9.10. A comparison between the properties of cyclones and anticyclones is of a fundamental interest. Such a comparison could answer two fundamental questions: (a) what is the origin of the cyclone-anticyclone asymmetry of large-scale, long-lived Rossby vortices, both in the laboratory and in the atmospheres of giant planets, and (b) whether the observed Rossby vortices have anything to do with solitons.

The preceding discussion implies that Rossby vortices of either polarity have the following features in common: first, they drift in the direction of Rossby wave propagation; second, with a large enough amplitude, they carry particles of the medium along in their drift; third, they collide inelastically and can, in particular, merge with one another.

Let us now consider the differences in the behavior of Rossby vortices of opposite polarity. Figures 9.11, 12 show experimental results concerning drift velocities and lifetimes of Rossby vortices [9.10]. The vortices are generated according to the techniques described in Sect. 6.2: the anticyclones are created with technique "c" (as a rule) or "b" (as an exception), the cyclones with technique "d". The developed vortices drift in the propagation direction of linear Rossby waves. In the experiments illustrated by Figs. 9.11, 12, the vessel rotation velocity Ω for all values of shallow water depth at the working parallel is chosen to

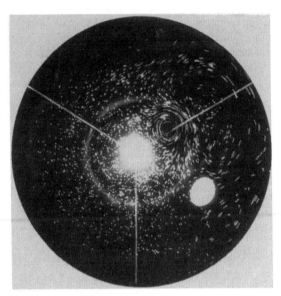

Fig. 9.10. A cyclonic Rossby vortex created with the pumping disk by technique "a" (the large paraboloid; due to an inward gradient of H_0, the cyclone drifts in the direction of vessel rotation, i.e., counter-clockwise)

be close to Ω_0,, i.e., to the nominal velocity corresponding to a liquid layer of uniform thickness $H_0 = 1$ cm. Under these conditions, the Rossby velocity V_R, according to (5.11), is the same (about 3 cm/s) for all values of H_0. [The conclusion that V_R is independent of H_0 for $\Omega = $ const can be drawn, for example, from (6.8): since the addition of more liquid shifts its equilibrium free surface by the same distance along the rotation axis at any point, the value of $\partial H_\eta / \partial y$ is not changed and V_R remains the same.]

The velocity profiles of cyclones and anticyclones created as described above are shown in Figs. 9.13, 14. Vortex amplitudes are $h_0 = 0.2$–0.4 for cyclones and $h_0 = 0.2$–1 for anticyclones. Larger values of h_0 correspond to smaller values of layer thickness H_0.

The ratio of vortex diameter (understood here as the distance $2a$ along the parallel between the points in the vortex where the linear velocity of its proper rotation is the highest) to the Rossby-Obukhov radius ranges from 2 to 4–5. The larger values of $2a/r_R$ correspond to $H_0 \approx 1.5$ cm, the smaller to $H_0 = 4$–5 cm.

The experiments indicate that the anticyclones, as they start drifting (while still having a relatively large amplitude), drift at a velocity which is usually somewhat higher than V_R – to be expected of solitons in accordance with (5.33). As the amplitude of an anticyclone decreases (due to viscosity), its drift velocity approaches from above the Rossby velocity $(\omega/k_x)_{\max}$, defined by the dispersion relation (5.8) as the maximum velocity of linear Rossby waves. The actual value of $(\omega/k_x)_{\max}$ is somewhat less than the idealized quantity V_R which corresponds to the limiting case $k_x, k_y \to 0$. The fact that $(\omega/k_x)_{\max}$ is lower than V_R may be

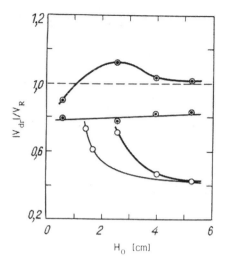

Fig. 9.11. Relative drift velocity V_{dr} of monopolar Rossby vortices as a function of unperturbed liquid layer thickness (the large paraboloid, $\Omega = \Omega_0$; *open circles* correspond to cyclones generated by technique "d", *filled circles* to anticyclones generated by technique "c"; *bold curve* corresponds to the values in the beginning of vortex life (higher amplitudes), *thin curve* to those in the end (lower amplitudes)

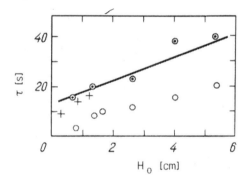

Fig. 9.12. Lifetime τ of monopolar Rossby vortices (the time required for the amplitude of linear velocity within the vortex to decrease by a factor of e) as a fuunction of unperturbed layer thickness H_0 (the large paraboloid, $\Omega = \Omega_0$). For anticyclones, the values of τ, averaged over a number of runs, are marked with *crosses* (technique "b") and *filled circles* (technique "c"); for cyclones, they are marked with *open circles* (technique "d"). The measured values of τ differ from the averages by certainly no more than 25%. The *straight line* is the viscous damping time τ_ν of a monopolar Rossby vortex, calculated according to (5.35)

due to the finite lengths of the longest actual waves in the x and y directions which can fit in the real vessel along its parallels and meridians. Thus the inequality $V_{dr} \gtrsim (\omega/k_x)_{max}$ mentioned above is not violated even though the drift velocity V_{dr} falls eventually to $0.8V_R$. Consequently, one can conclude that the observed anticyclones exhibit a property pertaining to Rossby solitons.

The drift velocity of the cyclones depends on how well they are shaped. With a relatively large liquid depth H_0 the cyclones are round and rather compact. They drift then significantly slower than the anticyclones (about twice as slow). The drift velocity of the cyclones, even in the end of their lifetime when the depression in the liquid surface vanishes, is close to $0.4V_R$. For small values of H_0 (say, about 1.5 cm), when the sizes of the vortices of either polarity expressed in the units of r_R grow significantly, a noticeable increase in cyclone drift velocity is observed (the drift velocity of anticyclones remains virtually the same). This phenomenon is similar to what is characteristic for linear packets of Rossby waves whose drift velocity also rises significantly with their size (Fig. 5.3). The different drift behavior of cyclones and anticyclones thus seems to be of a qualitative nature. It is one of the manifestations of the cyclone-anticyclone asymmetry exhibited by Rossby vortices.

Another manifestation of this asymmetry is revealed by comparing the lifetimes of the vortices of opposite polarities (Fig. 9.12). [Here we understand the lifetime of a free vortex as the interval during which the maximum value $(v_\varrho)_0$ of the meridional component of linear velocity in the vortex (the maximum linear rotation velocity across the profile passing through a parallel of the vessel) is reduced by a factor of e. This time is obtained from the linear dependencies $\ln(v_\varrho)_0 = f(t)$ observed in the experiments. The linear character of these dependencies implies, in particular, that the difference between the lifetimes of cyclones and anticyclones is not due to their possibly different amplitudes.] One can see that anticyclones live much longer than cyclones, especially in the range of relatively small liquid depths ($H_0 \lesssim 1$ cm). This qualitative distinction in the behavior of opposite-polarity Rossby vortices was noticed already in our first experiments [9.2, 5]. We believe it to be fundamental and interpret it as follows. The conditions for the generation of cyclones and anticyclones are such that the absolute dimensions of the vortices – at the initial moment of their creation – are approximately the same. Therefore the ratio of vortex size to the Rossby-Obukhov radius grows substantially as H_0 is decreased. Then, as the experiments show, anticyclones retain their round shape while cyclones become greatly deformed, stretching along the parallel and much growing in size. We are inclined to explain this difference in behavior to the lack of balance between dispersion and scalar nonlinearity in the cyclones. The latter nonlinearity, under these conditions, contributes to cyclone decay, which naturally is manifested stronger for higher values of the ratio a/r_R, i.e., when the inequality (5.21) is stronger. It is therefore no wonder that the difference in the lifetimes of cyclones and anticyclones is the highest for small H_0. This is yet another manifestation of the cyclone-anticyclone asymmetry.

Note that Fig. 9.12 also shows the measured lifetimes of anticyclones produced according to technique "b" by injecting a jet of liquid along vessel parallel, for the range 0.4 cm$\leq H_0 \leq 1.2$ cm. These results are in a good agreement with those obtained according to technique "c" and described above.

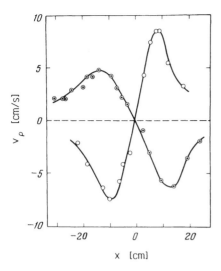

Fig. 9.13. Spatial profiles of proper rotation of monopolar Rossby vortices with a relatively thick liquid layer ($H_0 = 5$ cm; the large paraboloid): meridional component v_ϱ of linear velocity as a function of the coordinate x measured from vortex center along the parallel in the direction of vessel rotation, for an anticyclone (*filled circles*, technique "c") and a cyclone (*open circles*, technique "d")

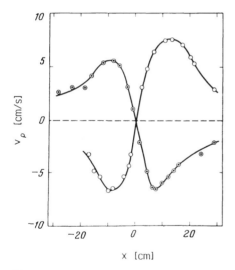

Fig. 9.14. Same as in Fig. 9.13 but with a relatively thin liquid layer ($H_0 = 1.5$ cm)

It is easy to see from the data shown in Fig. 9.12 that, over the entire range of H_0, an anticyclone whose maximum layer thickness is lower than the minimum thickness in a cyclone (with due account for either an elevation or a depression of the free surface of liquid in the vortices) nevertheless lives considerably longer than the cyclone. This is an additional argument for the viewpoint that the shorter lifetimes of cyclones (as compared to anticyclones) cannot be explained simply by their smaller thickness and must be due to other, more complicated reasons discussed above.

A qualitative difference in the structure and lifetimes of large cyclones and anticyclones – in full accordance with the results described here – is also revealed in numerical calculations [9.28–31]. A similar cyclone-anticyclone asymmetry is observed in the atmospheres of Jupiter and Saturn [9.26, 27, 32–35]. The following general rule is found to be valid there: the largest and longest-lived vortices are all anticyclones, their sizes exceeding the inner Rossby radius r_i – which means that they correspond to the regime of (5.21). Long-lived cyclones are rare and may be regarded as exceptions. Such are the chains of Barges in Jovian atmosphere (14°N) and the relatively small vortex UV Spot in Saturnian atmosphere (24°N) [9.27, 36]. In both cases, the characteristic (meridional) size of the cyclones is smaller than the Rossby radius (or close to it), which corresponds better to the regime of (5.22).

Let us now return to Fig. 9.12. For comparatively deep liquid layers ($H_0 > 1$ cm), the difference in the lifetimes of anticyclones and cyclones is not as high as for $H_0 \lesssim 1$ cm. However, it is still high enough to be a definite indication of the fundamentally different behavior of cyclones and anticyclones. Let us consider this range in more detail. Here the lifetime ratio of anticyclones and cyclones is $\zeta = \tau_a/\tau_c \approx 2.3$. The important point is that the value of ζ is considerably larger than the ratio of the respective characteristic viscous damping times $\tau_{\nu a}$ and $\tau_{\nu c}$. Indeed, it is easy to show (using the explanations given in Sect. 5.6 and the relations (5.34, 35) in particular) that $\tau_{\nu a}/\tau_{\nu c} \lesssim 1.3$ for the observed vortices of approximately equal profiles and amplitudes (from here on, we shall assume $\tau_{\nu a}/\tau_{\nu c} = 1$ with no detriment to our discussion). Based on this data, the following qualitative estimates can be made for the dispersion spreading times of cyclones and anticyclones. Since the rates of the processes leading to vortex decay are additive (this was shown numerically by A.V. Khutoretsky [9.10]),

$$1/\tau_c = 1/\tau_{dc} + 1/\tau_{\nu c} , \quad 1/\tau_a = 1/\tau_{da} + 1/\tau_{\nu a} \tag{9.1}$$

where τ_{dc} and τ_{da} are the dispersion spreading times of cyclones and anticyclones, respectively.

The compatibility condition for (9.1) yields (with $\tau_{\nu a} = \tau_{\nu c} = \tau_\nu$)

$$\frac{\tau_{da}}{\tau_{dc}} = \frac{\zeta}{1 - (\zeta - 1)\tau_{dc}/\tau_\nu} . \tag{9.2}$$

And since the left-hand side of (9.2) cannot be negative, the cyclone dispersion time is limited by

$$(\tau_{dc})_{max} = (\zeta - 1)^{-1} \tau_\nu \tag{9.3}$$

or, using the values corresponding to the conditions of our experiments with $H_0 > 1$ cm,

$$(\tau_{dc})_{\max} \approx \tau_\nu \approx \tau_1 \,.$$

But the result we have obtained, $(\tau_{dc})_{\max} \approx \tau_1$, means that cyclones undergo a dispersion spreading with the characteristic time of a linear Rossby wave packet on the β-plane, and the condition (5.15) is not valid for them. Thus Rossby cyclones are not solitons. This conclusion also agrees with the fact that the observed free-run length of even the most "long-lived" cyclones during their typical lifetime does not exceed one and a half proper diameters of the cyclone, i.e., it proves to be virtually the same as that of a linear Rossby wave packet of the same size, Fig. 5.3. It is easy to see that the free-run length of an anticyclone (limited by viscosity) is about five times greater.

This result is confirmed by the numerical calculation conducted in [9.24, 25]. It is shown there that if the size of a cyclone is significantly larger than r_i and its proper rotation velocity V_r is not too high (e.g., if it is only several times higher than the drift velocity as in our experiments) then the cyclone lifetime is no greater than that of a linear Rossby wave packet of the same size (Fig. 5.3). Only in the case of $a \lesssim r_i$ and $V_r \gg V_{dr}$ can the vector nonlinearity prolong significantly the life of a cyclone. The latter effect is apparently responsible for the persistence of the long-lived cyclonic Brown Ovals of Jupiter and some of the cyclonic rings of the Gulf Stream (see Sects. 2.2, 2.3.2, 10.3).

The estimate of cyclone lifetimes given here makes it possible to draw an interesting qualitative conclusion concerning anticyclones, too. Indeed, assuming cyclone dispersion time to be equal to its maximum possible value (9.3), we see from (9.2) that anticyclone dispersion time then becomes arbitrarily long. This means that the condition (5.15) is valid and the anticyclones in question are Rossby solitons.

There is also a direct experimental proof of this conclusion. As we have already noticed in Sect. 5.6, the characteristic viscous damping time of Rossby vortices, according to (5.35), must tend to a *finite* limit as $H_0 \to 0$. When applied to the conditions of our experiments, the theoretical dependence $\tau_\nu(H_0)$ looks like the solid line in Fig. 9.12 (see [9.10]). It is seen from Fig. 9.12 that this curve virtually coincides, both qualitatively and quantitavely, with the dependence observed experimentally for anticyclones. Consequently, anticyclone lifetime is practically entirely determined by viscosity. (As to the cyclones, they decay, as we have seen, in times much shorter than the viscous time if H_0 is small. According to our interpretation, their decay is related to dispersion spreading.)

The following interesting fact is worth mentioning: it is easy to create such conditions (using either technique "b" or technique "c") that an anticyclonic vortex is produced whose elevation with respect to shallow water surface $(\delta H)_0$ is much greater than the initial liquid depth H_0. Then, formally, $h_0 = (\delta H)_0/H_0 > 1$. The experiments show that such "exotic" vortices have about the same lifetimes as those with $h_0 < 1$. This is true even for very shallow layers, down to $H_0 \approx 0.1$ cm. These facts are another indication of the "ruggedness" of the solitonic structures in question.

It is also interesting to note the similarity of the large-amplitude ($h_0 \gtrsim 1$) solitary anticyclones to anticyclonic intrathermoclinic vortices – lenses, observed in the oceans (Sects. 2.3.3, 10.3). The thickness of such lenses falls off to zero at their edges, making them similar to the anticyclones with $h_0 \gtrsim 1$ mentioned above.

Discussing the problem of cyclone-anticyclone asymmetry, two more facts should be pointed out. The first of them, observed in our experiments and described in Chap. 11 (Fig. 11.6), is that an initially created "chessboard" pattern containing vortices of both polarities evolves gradually, after the pumping has been switched off, into a set where only anticyclones persist. The second result, obtained numerically for shallow water in the β-plane approximation [9.37], is that an initially introduced "polarity-symmetrical" vortex system, containing equal numbers of interacting cyclones and anticyclones of the same amplitude and size ($a \approx 2r_R$), evolves into another – stationary – state characterized by a sharp cyclone-anticyclone asymmetry. In the latter state, several anticyclones clearly are predominant vortices: their sizes are much larger than initially ($a \approx 4r_R \approx r_{IG}$) and their amplitude corresponds to a very high nonlinearity: $h_0 \to 1$. As to the cyclones, they become fractured ($a \approx r_R$) and considerably weaker than they were initially. In the case when initial vortex dimensions exceed $4r_R$, system evolution leads to diminishing vortex scales, and most of the energy is again concentrated in anticyclones whose size is $a \approx 4r_R \approx r_{IG}$. Thus anticyclones of size about r_{IG} exhibit attractor properties. This phenomenon is unambiguously interpreted by the authors of [9.37] as the effect of the scalar nonlinearity whose influence is dominating in the range $a > r_R$. It is important to point out that the limiting vortex size ($4r_R$) has proved to be equal to the intermediate geostrophic radius (5.32). These numerical results are in excellent agreement with the laboratory experiments described here.

9.5 Quasi-Two-Dimensionality of Rossby Vortices. The Non-Principal Role of Viscosity

The bulk of experimental evidence presented here allows one to treat the Rossby vortices we are studying as practically two-dimensional. This is supported by the following arguments. The thickness of the bottom Eckman layer (4.5) in our experiments is $l_E \approx 0.4$ mm. The depth of the liquid is varied from several millimeters to over 50 mm. Thus, shallow water depth is always much greater than the thickness of the Eckman layer (from seven times to more than two orders of magnitude). In other words, liquid motion is virtually uniform over the entire main layer of the vortex and therefore may well be considered two-dimensional. This is illustrated by Figs. 9.11, 12 as well as Fig. 11.5 to be discussed later, which testify that all the properties and behavior of the Rossby vortices under investigation, in particular the ratio of vortex drift velocity to the Rossby velocity V_R, are independent of liquid depth H_0 across a wide range of its values. This is natural because the Eckman number in the above experiments ranges from $E \lesssim 10^{-4}$ to $E \approx 10^{-2}$. (The experiments described in [9.38, 39] are consistent with ours. They show that the main laws of Rossby vortex generation do not depend on layer thickness in a rather wide range, from 10 to 60 Eckman layers, even though the zonal flows are created quite differently.)

In other words, the experiments indicate that the viscosity of the medium does not significantly affect the properties of the vortices in question. Accordingly, direct experiments with $NiSO_4$ solutions of different concentrations show that even for smallest layer thickness of about ten Eckman layers ($H_0 = 0.3$–0.5 cm), the dynamics of Rossby vortices described here remains the same if the viscosity of the medium is increased approximately threefold; consequently, viscosity affects only vortex lifetime. This is natural since the time of vortex viscous damping (practically equal to the observed lifetime of anticyclones under the conditions of Fig. 9.12) is much greater than the typical period of vortex rotation: e.g., it is about an order of magnitude greater for $H_0 \approx 5$ cm.

The measurements of the lifetimes of Rossby anticyclones appear to provide the most convincing proof of their two-dimensionality. Indeed, as shown in Fig. 9.12, the measured lifetime of anticyclones as a function of layer thickness H_0 is in a very good agreement, both qualitative and quantitative, with the dependence $\tau_\nu(H_0)$ predicted by the two-dimensional theory describing the viscous damping of Rossby vortices and given by (5.35). In particular, as $H_0 \to 0$, the measured lifetime of anticyclones tends to a finite (and high enough) limit rather than to zero in full accordance with that theory. Therefore the anticyclonic Rossby solitons in question can be treated as two-dimensional objects.

This conclusion is confirmed by the result of an experiment carried out as a continuation of those described in [9.10]. In the experiment, the so-called "spin-down" time τ_{sd} is measured, i.e., the duration of the transition process in which the liquid in the rotating vessel reaches a new stationary state after an abrupt decrease in vessel rotation velocity by a small amount (a similar process corresponding to an abrupt increase in the rotation velocity is called "spin-up"). The characteristic time of this process is known to be equal to the characteristic viscous time [9.40]. The experiment yields the result shown in Fig. 9.15. One's attention is drawn to its qualitative similarity to what is shown in Fig. 9.12, that is: first, the linearity of the dependence $\tau_{sd}(H_0)$; second, the rather high nonzero value of τ_{sd} for $H_0 \to 0$. In the light of the above interpretation of Fig. 9.12, the result shown in Fig. 9.15 appears quite natural: it cannot be different if the liquid has a free surface. The difference between the conditions of Figs. 9.12 and 9.15 is in the value of the Froude number $F = l^2/r_R^2$ (see (5.35)): in the first case $l = a$ (where a is vortex size), while in the second case l is the characteristic radial dimension of the vessel, i.e., a quantity about twice greater. This is the primary explanation for the quantitative difference between these two figures. A similar result has been obtained in the experiments reported in [9.41], showing that for a free-surface shallow water the spin-down time at high values of the Froude number is proportional to $\tau_E F$, i.e., does not depend on H_0.

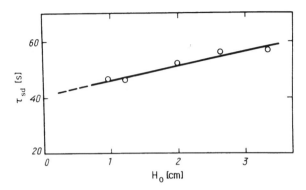

Fig. 9.15. Characteristic spindown time τ_{sd} as a function of unperturbed liquid layer thickness H_0. The large paraboloid; vessel angular rate $\Omega + \Delta\Omega$ in the initial state and $\Omega - \Delta\Omega$ in the final state, $\Delta\Omega \approx 0.03\Omega$; H_0 is measured at the parallel $\varrho = D/2\sqrt{2}$ where $\partial H_0/\partial y = 0$ in the case of equilibrium rotation at angular rate Ω

Thus, since the viscous time of the objects under investigation agrees well with the two-dimensional theory, the experiments cited above testify to the two-dimensionality of the Rossby vortices in question.

The experimental data discussed in this section are of a crucial importance for laboratory simulation of natural quasi-two-dimensional Rossby vortices. They also demonstrate a very clear and fundamental fact: with a free surface present, the viscous damping time of Rossby vortices τ_ν is quite different from the Eckman time τ_E, see (5.35). The understanding of this fact helps explain the seemingly paradoxical result shown in Fig. 9.12, namely, that the time τ_ν does not depend on H_0 for small values of H_0.

To conclude this section, a quite important remark must be made. As a matter of fact, our experiments show that in order to produce solitary vortices in thick liquid layers ($H_0 \gtrsim 1.5$ cm), one should use techniques "c" and "d". As regards techniques "a" and "b", under such conditions they generate vortex pairs, i.e., dipolar rather than solitary vortices. (The same feature with respect to technique "b" is also pointed out in [9.42], to be discussed in Chap. 11.) Solitary vortices are formed later, *after* one of the partners in the pair has decayed. Usually, it is the cyclone that decays sooner and the long-lived anticyclone is left. If, however, the anticyclone in a vortex pair is loosely shaped or if it brushes against vessel wall or lid, then most of the energy of the dipole will be eventually concentrated in the cyclone. But in this case the produced cyclone is not a free vortex – it is fed by the nonstationary (for the above reasons) anticyclone. Therefore, in order to generate free monopolar vortices, techniques "c" and "d" must be used. This should be taken into consideration by those who would like to reproduce the experimental results shown in Fig. 9.12.

9.6 Vortex-Wave Dualism

The parameters of the vortices illustrated in Figs. 9.1–14 and the features of their propagation – namely, their dimensions, which clearly are related to the Rossby-Obukhov radius r_R, the direction and velocity of their drift, their cyclone-anticyclone asymmetry (the anticyclones are long-lived while the cyclones are not) – correspond well to the dispersion equation (5.8) for Rossby *waves*, if the qualitatively distinct nonlinear effects discussed above, such as the KdV nonlinearity, are taken into account. Consequently, the vortices we are studying possess conspicuous wave properties, i.e., their behavior substantially depends on the dispersion of Rossby waves. The cyclones behave like spreading linear wave packets and the anticyclones like nonspreading solitons.

The strong influence of dispersion on the properties of these Rossby vortices can also be seen from other facts. For instance, according to (7.13), the dispersion of Rossby waves can be counterbalanced by a uniform flow with a velocity $u = -V_R$ directed westwards, along the propagation of Rossby waves. (Recall that the positive direction for u is eastwards and that V_R is the velocity magnitude.) In this case, $\beta(u) = \beta(0) - V_R r_R^{-2} = 0$, i.e., the dispersion is canceled out. But then, according to our interpretation, the cyclone-anticyclone asymmetry must also vanish. This phenomenon is indeed observed in a detailed numerical calculation [9.28–31] which shows that in a westward flow with velocity $u = -V_R$ vortices of either polarity suffer no dispersion and their lifetimes are equal.

Another fact, noticed in laboratory simulations, in numerical calculations, and in observations of the JGRS, that the preferable size of long-lived Rossby vortices is the intermediate geostrophic radius r_{IG}, i.e., the maximum size of a stationary monopolar Rossby soliton which is immune from spreading caused by the spatial nonuniformity of the Rossby velocity – also points to a strong influence of dispersion on the properties of such vortices.

On the other hand, the structures under investigation are "genuine" vortices: as they drift, they carry along liquid particles enclosed within a particle capture region (inside the separatrix). For example, it can be seen from Fig. 9.4 that the trapped particles drift with the vortex with respect to the vessel while those outside of the separatrix lag behind. The experiments show that a particle capture region exists in a Rossby vortex only if (2.15) is satisfied, i.e., if the maximum linear velocity across the vortex rotation profile exceeds the velocity of vortex drift. This condition is satisfied in all of our experiments described here. Therefore, many structural and dynamic features of the objects we are studying are related to their vortical properties. Moreover, it turns out that not only the specific behavior of these structures – the Rossby solitons – but the very fact of their existence is due to the presence of captured particles in them. This follows both from the experiments that have been described and from their comparison with the theory (Sect. 9.7).

The interdependence between the wave and vortical properties of Rossby solitons, indicated above, is a demonstration of their vortex-wave dualism. So the structures we investigate may be termed both vortical Rossby solitons and Rossby wave vortices. We shall see other manifestations of this dualism when we discuss dipolar Rossby vortices (Chaps. 11, 12).

9.7 Comparison between Experiment and Theory

The experimental data discussed above made it possible to identify the observed anticyclones as Rossby solitons and to establish that the cyclones are not solitons. In this identification, we have relied to a great extent on qualitative conclusions drawn from the theory which treats Rossby solitons as solitary waves. The wave concept directed us to search for solitons among Rossby structures of a more or less definite spatial scale – with characteristic dimensions greater than the Rossby-Obukhov radius. It helped understand the laws of their drift and the cause of their cyclone-anticyclone asymmetry. Thus the wave concept played an important euristic role in our experiments.

However, our experiments have also revealed that the purely wave concept of Rossby solitons is in a crucial contradiction with quite a number of experimental facts.

1. A Rossby soliton as a pure solitary wave, according to the theory, must have the size given by (5.30), and this size must be inversely proportional to the square root of its amplitude. The experiment, however, contradicts this theoretical prediction. Indeed, Fig. 9.3a implies already that a change in soliton amplitude by more than 2.5 times (with the other conditions of the experiment unchanged) does not tell in any way on the width of its profile. Even more spectacular is Fig. 9.16 where the observed soliton profiles are shown together with the theoretical profiles given by (5.29, 30). The contradiction with the theory is apparent: the experimental profile of a Rossby soliton and its characteristic size are independent of its amplitude.

2. The wave theory presumes that particles are not involved in wave propagation, which clearly contradicts the observed particle transport in the drifting Rossby solitons (Fig. 9.4).

3. According to the theory, solitary waves must pass through each other freely upon collision – to put it more precisely, they must collide elastically. This property of a solitary wave is even included in the definition of a soliton adopted in mathematics and mathematical physics [9.43–49]. Physicists, however, often use the definition which identifies wave solitons with solitary waves, regardless of their behavior in collisions [9.50–55]. We also prefer this definition (Sect. 5.4). Therefore the fact that Rossby anticyclones collide inelastically (Fig. 9.9) does not prevent us from calling them solitons. However, it contradicts the wave concept.

4. Finally, the purely wave concept contradicts the more rigorous theory given in Sect. 5.5 which takes into account the dualism of Rossby solitons, i.e., the fact that they possess not only wave but also vortical properties. The latter, more

comprehensive, theory implies, as was pointed out, that a Rossby soliton must be a genuine vortex, i.e., it must contain trapped particles drifting along with it. This conclusion agrees well with the experimental data given in Sect. 9.2 and Fig. 9.4.

A major step in the development of the theory has been made in [9.19–25]. In these studies, accomplished by numerical calculation, only anticyclonic Rossby vortices were analyzed which have sufficiently large amplitude and carry trapped particles with them. The vortices were also assumed to be axially symmetrical and much larger in size than the Rossby-Obukhov radius. The vector nonlinearity was therefore considered negligibly weak, and the principal nonlinear factor was assumed to be the scalar (KdV) nonlinearity. This approach yielded the following results.

The effective size of an anticyclonic Rossby soliton is independent of its amplitude; it is determined simply by the size of the region where, according to the conditions of vortex generation, its amplitude is large enough to satisfy the particle capture condition (2.15). It is easy to see from Fig. 9.16 that the calculated profiles of such vortical solitons "remembering" the conditions of their generation practically coincide with the measured ones (see also [9.56]).

This property of a vortex with closed streamlines, its "memory", pointed out in the new theory, is explained by the fact that any deformation of the vortex is strongly hampered if it is rotating fast enough [9.25]. In particular, vortex deformations that lead to Rossby wave radiation are hampered. This is essentially the same nonlinear effect which slows down the dispersion decay of rapidly rotating cyclonic Gulf Stream rings. This effect is obviously directly connected with the Larichev theorem [9.57] on adiabatic invariance of closed potential vorticity isolines (we mean here the invariance over time intervals which are small compared to the viscous time). As was already pointed out, this effect vanishes when the vortex amplitude is not large enough – when the maximum linear velocity in the vortex is not greater than its drift velocity.

A vortical soliton of this kind [9.19, 20] is typically not analytic, i.e., the higher-order spatial derivatives of its vorticity are discontinuous at its separatrix in the framework of the idealized model (with no dissipation). This discontinuity is, however, rather of a purely theoretical nature, since in a real medium it is smoothed out due to the inevitable viscosity. There are many such "nonanalytical" solitons for a given size, so they have a much greater "statistical weight" than the unique (for the same size) analytical soliton which, depending on the size, is either a genuine vortex with trapped particles [9.25] if its amplitude is high or a solitary wave (5.29, 30) with no trapped particles if it is low. Accordingly, the nonanalytical vortical solitons should appear with much higher probability than the analytical ones. This is confirmed by the experimental data presented in Figs. 9.1–14. The solitons shown in these figures exhibit properties which are in a good agreement with those of the vortical solitons described here. The theory [9.25] perfectly allows that under some specially chosen conditions the high-amplitude analytical Rossby soliton containing trapped particles can be obtained.

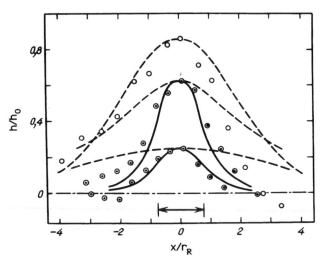

Fig. 9.16. Experimental profiles of free surface elevation in Rossby anticyclones compared with the theoretical curves for different amplitudes: *filled circles* are the data from Fig. 9.3a, *open circles* are those from Fig. 9.3b, *solid curves* represent the numerical calculation [9.19] for a vortex containing trapped particles in its central region (marked with *arrows*) with a given constant potential vorticity (the diameter of particle trapping region $1.5r_R$ is equal to that of the pumping disk); *dashed curves* are the particular (smooth) analytical solution [9.19–21], practically coinciding with (5.29, 30)

It is interesting that the profile of this soliton also is practically described by (5.29, 30) up to the maximum possible value $h_0 \approx 1$. We must note in this connection that in the experiment a soliton of a near-unity amplitude ($h_0 \approx 0.9$; Fig. 9.3b,c) has a profile which is quite comparable to that defined by (5.29, 30), see Fig. 9.16.

The theory [9.24, 25, 58] comes to the conclusion that the vortical property of the solitons in question is a prerequisite for their existence: a Rossby anticyclone whose amplitude is not enough for it to contain trapped particles decays relatively quickly – practically in the characteristic dispersion spreading time of a linear wave packet of the same size, i.e., it is not a soliton. The phenomenon responsible for this decay is related to the variation of the Rossby velocity along the meridional coordinate and is taken into account by the last term in (5.4). It exists only in the framework of the "purely wave" concept. If, however, the vortex amplitude is large enough (such that the characteristic velocity of particle rotation in the vortex, proportional to it, is considerably higher than the velocity of vortex drift) then this effect vanishes as a consequence of the averaging of particle motions over the "northern" and the "southern" parts of the vortex. (It is easy to see from the condition of geostrophic equilibrium (2.11) that the rotation in the vortex is sufficiently fast if $h_0 > a/R$.) Since in the laboratory experiments no such effect is observed for anticyclones of large enough amplitudes, this means that the experiment testifies against the purely wave concept.

In connection with the phenomenon we are discussing, the following fact which had manifested itself clearly in the very first experiments on Rossby vortices should be recalled: the vortices (even anticyclones) decay quickly if their sizes are too large, e.g., such that $a \gtrsim 3r_R$ (Sect. 9.2). The limiting size of the vortices proved to be virtually equal to the intermediate geostrophic radius (5.32), see also (5.31) and the comments on that relation.

It is possible that the above effect is what stops the so-called "inverse cascade" in which vortices merge and enlarge, leading to the formation of broad zonal flows in planetary atmospheres [9.59, 60]. There are indications to the possibility that the inverse cascade is stopped by this effect both in theory [9.61] and in numerical calculation [9.37]. As a matter of fact, according to [9.37], a system of merging geostrophic vortices can have a stationary state in which the maximum size of the structures equals the intermediate geostrophic radius r_{IG}. So, if flow width, which is half the Rhines length $l_{Rh} \approx \pi(2u/\beta)^{1/2}$, is substantially larger than r_{IG}, then the cascade may not reach the zonal flow phase. It is interesting that in the atmospheres of the giant planets $r_i \approx 10^3$ km, $r_{IG} \approx 4 \cdot 10^3$ km $\approx l_{Rh}/2$, so the formation of zonal flows does not contradict the result obtained in [9.37].

According to the theory [9.24], collisions between vortical solitons lead to their merging, in full agreement with our experiments (Fig. 9.9). This result, obtained in [9.24], is also confirmed by the numerical computations carried out by other authors [9.29–31]. The merging (i.e., reconnection of vortex streamlines) requires, in principle, some degree of dissipation, but this is always present in reality – both in laboratory experiments and in numerical calculations (where, in order to ensure computational stability, the so-called computer viscosity is introduced). In the numerical calculation described in [9.62], however, vortex merging occurs even with no dissipation but in the presence of small-scale turbulence which also can appear under such conditions.

The theoretical drift velocity of the solitons in question is higher than the Rossby velocity V_R, albeit by a relatively small amount: e.g., $V_{dr} \approx 1.2V_R$ for $h_0 \approx 0.6$ [9.20]. This is also close to the experimental data (Fig. 9.11).

Finally, only anticyclones possess solitonic properties. Cyclones (whose radius is greater than r_R) live little longer than the lifetime of a linear Rossby wave packet. This result of the theory [9.25] is also in a good agreement with the experimental data described above.

Thus we may assert that the whole bulk of experimental evidence obtained in our laboratory simulation agrees well with the theory of vortical Rossby solitons discussed here. This agreement, as can be seen from Fig. 9.12, also extends to the behavior of anticyclones with respect to viscous damping.

10. Solitonic Model of Natural Vortices

This chapter describes a model of long-lived, large-scale Rossby vortices in planetary atmospheres and in the oceans. In spite of all the distinctions between such vortices with respect to the ambient media and to their dimensions (thousands of kilometers in atmospheres and dozens of kilometers in the oceans), they have the following major features in common: (1) they possess a clearly manifested cyclone-anticyclone asymmetry – all of them, with little exception, are anticyclones; and (2) their sizes are somewhat greater than the Rossby radius r_i. These and other properties of such vortices make it possible to describe them with a model which is based on the theory of vortical Rossby solitons and discussed in the next sections.

10.1 Solitonic Model of the JGRS and Other Large-Scale, Long-Lived Vortices in Planetary Atmospheres

The views on the physical nature of such vortices as the JGRS have a history of their own. Here is a brief account of the major milestones. The author of [10.1] interpreted the JGRS vortex on the basis of the "Taylor column", a beautiful hydrodynamic phenomenon [10.2] which is derived directly from the Rossby-Ertel theorem on the conservation of potential vorticity. Suppose that a flow exists in a rotating shallow water over a solid bottom which has a topographic peculiarity, a "stump" (that is, a place where the liquid layer thickness H_0 is less than elsewhere). Then, by virtue of the above-mentioned theorem (5.13), the total curl of the flow (that is, rot $v + f$) will be smaller over the "stump" than in the surrounding regions, giving rise to an anticyclonic vortex at this place (the sign of the first term in the numerator is opposite to that of the second one). According to [10.1], the anticyclonic vortex of the JGRS is related to such a topographic peculiarity. However, it is now well known that the JGRS vortex is *drifting* relative to the planet (making a complete revolution along its parallel in 4–5 years). So today the hypothesis proposed in [10.1] can be only of an historical interest.

In [10.3], using a number of simplifying assumptions, it was shown by the numerical method that the well-known equation describing a vortex in a system of atmospheric zonal counterflows leads to the conclusion that, even without an underlying surface, there can be a vortex whose characteristic size is substantially greater than the Rossby-Obukhov radius (which is true for the JGRS).

The study reported in [10.3] was already a close antecedent of the theoretical solitonic model of the JGRS where that vortex is treated as a Rossby soliton. The first consistent theory of a solitonic Rossby vortex existing on the background of zonal counterflows was discussed in [10.4-7]. The primary nonlinear factor which counterbalances the dispersion of Rossby waves, making possible the formation of the soliton, was assumed to be the vector nonlinearity. The necessary condition for the manifestation of this nonlinearity is the absence of axial symmetry, since otherwise the Jacobian in (5.1) becomes zero. The JGRS vortex, in particular, satisfies this condition, having the shape of an oval, the ratio of the major axis (along the parallel) to its minor one being about 2:1. With its nearly steady propagation with respect to the planet and the lack of any apparent degradation over three centuries, it clearly resembles the soliton which was the subject of the theory presented in [10.4-7].

In addition to the zonal flows wrapping the JGRS, the theory [10.4-7] took into account the vertical inhomogeneity of the atmosphere. That theory, developed in 1976, had at that time an indisputable methodological advantage: it showed that, within the approximation where one of vortex dimensions (along the parallel) is assumed to be much larger than the other (along the meridian), the nonlinear equation of the vortex becomes almost one-dimensional and looks like the KdV equation. It then implies, similar to a solution of the KdV equation, the existence of an anticyclonic Rossby soliton in anticyclonic shear flow which resembles the JGRS, albeit quite grossly.

The theory of [10.4-7] did not include the scalar nonlinearity (the latter was discovered only afterwards). This was a major drawback of that theory, because in the case of the JGRS the condition (5.22) for the vector nonlinearity to prevail is obviously invalid; on the contrary, the condition (5.21) for prevailing scalar nonlinearity is satisfied. Therefore the theory given in [10.4-7] and, of course, the preceding study reported in [10.3] could not adequately explain all the major features of the JGRS vortex. In particular, the theory of [10.4-7] predicted too high vortex drift velocity – an order of magnitude higher than the actual value [10.8]. A major shortcoming was that it was a purely wave theory, treating the vortex in question as a solitary wave propagating separately from the particles. It predicted therefore that collisions of the JGRS vortex with other large vortices must be elastic: the vortices should pass freely through each other. The observations described in [10.9-11] have refuted this prediction: they have shown that such collisions are actually inelastic and lead to the merging of the partners.

A modified solitonic model of the JGRS, also based on the vector nonlinearity and taking account of the zonal flows, but two-dimensional as contrasted with [10.4-7], was presented in [10.12, 13]. It is important to note that in [10.12, 13] the natural object JGRS is treated not as a wave soliton proposed in [10.4-7] but as a genuine vortex with closed streamlines. This is a crucial fact which has a profound physical meaning (see the previous chapter).

A scalar two-dimensional analytical solitonic model treating the JGRS as a solitary vortex in a geostrophic flow with a velocity shear was discussed in [10.14].

Both vector and scalar nonlinearities – in particular, as applied to the JGRS – were simultaneously taken into account in the detailed numerical studies published in [10.15–18], based on the analysis of (5.4). The results of [10.15–18] also demonstrate the cyclone-anticyclone asymmetry discussed above: only anticyclones are non-spreading structures (that is, solitons). According to the numerical calculations of [10.17], they live (in anticyclonic shear flows) for more than 100 years, while cyclones (under the same conditions in cyclonic shear flows) live no longer than several months, that is, approximately the same as linear wave packets. (The authors of [10.16–18], using the term "solitary vortex", do not refer to their model as solitonic and to the long-lived vortices as solitons, but this does not affect the essense of the matter.)

The combined model [10.16–18] explains such features of the natural vortex in question as its size (exceeding r_R), its westward drift and anticyclonic vorticity, as well as its capability to merge with other vortices of the same polarity which is observed in many instances [10.9–11]. It is found that, in order to achieve a good quantitative agreement between the theory and the observations, it is sufficient to treat the JGRS and the surrounding zonal flows (in theory) as *two-dimensional*, taking into account the vertical inhomogeneity of the atmosphere in the same way, in principle, as we have done in Sect. 7.4 [10.19, 20]. Specifically, in [10.16–18] the replacement of the two-dimensional dispersion scale r_R by the three-dimensional scale r_i is done simply by replacing the acceleration of gravity g with the "corrected" acceleration $g' \approx g/2$, which, according to (7.9), is equivalent to using the value of $r_i \approx r_R/4$. This value of r_i is numerically close to our $r_i \approx r_R/6$ given by (2.10) above.

It was also shown in [10.16, 18] that, in the absence of dissipation, an anticyclonic vortex with the parameters of the JGRS is long-lived even with no zonal counterflows. In reality, however, the zonal flows are necessary, as we believe, to generate the vortex and to maintain its stationary state, in particular – to compensate for the viscous (and maybe other) losses of momentum. Attention was drawn to the important fact that the JGRS has just those dimensions, close to the intermediate geostrophic radius r_{IG}, which are to be expected for a Rossby soliton, namely:

$$r_i \ll a \lesssim r_i(R/r_i)^{1/3} = r_{IG} \tag{10.1}$$

where, in contrast to (5.31, 32), r_i is used instead of r_R. It should be noted that, of all the atmospheric vortices, the JGRS seems to be the only one to satisfy this condition. In particular, the terrestrial atmosphere has not enough room for that, so there is no wonder that the terrestrial atmospheric vortices are not solitons.

There are also theoretical models of a *free* Rossby soliton, propagating in the absence of zonal flows (assuming zero viscosity). They are based on the scalar nonlinearity. One of them [10.21] has been specially constructed for the synoptic vortices in the oceans (with due replacement $r_R \rightarrow r_i$). The other [10.22, 23] is related to the JGRS; it is strictly two-dimensional and can be shown to fit the observational data only when the replacement $r_R \rightarrow r_i$ is made (see [10.19, 20, 24] and Sect. 7.4).

We have enumerated the theoretical models of the JGRS. All of them, at the best, treat the vortex in question as existing in a *stationary* inviscid flow with a velocity shear, and leave two crucial questions unanswered: (a) how the vortex is generated by the zonal flow? and (b) why the JGRS vortex is unique over the entire planet perimeter? The answers to these questions have been given on the basis of the laboratory solitonic model of the JGRS (and other, physically similar, vortices) constructed in our simulation experiments and described in Chap. 7 and Sect. 8.3.

The stationary version of the laboratory model of the JGRS, where a Rossby soliton, unique over the entire planet perimeter, is continuously generated by zonal counterflows, is demonstrated in Figs. 7.6, 8. The nonstationary version of the model, where the Rossby soliton propagates freely in the absence of zonal flows, is shown in Figs. 9.1, 4. Figure 9.9 shows the merging of Rossby solitons observed in our experiments, which is also characteristic of the simulated long-lived vortices of Jupiter and Saturn. The laboratory model also solves the crucial problem about the physical nature of the cyclone-anticyclone asymmetry observed in the atmospheres of the giant planets (as well as in some kinds of oceanic vortices).

If we are to treat the long-lived atmospheric vortices as solitons, it would be interesting to compare their actual lifetimes with the minimum lifetime of a linear Rossby wave packet. Table 2.1 gives $(\tau_l)_{min} \approx 20$ terrestrial days which is approximately $5 \cdot 10^3$ times smaller than the time span over which the JGRS is observed.

Yet another argument supporting the conclusion that the JGRS vortex is a Rossby soliton deserves mentioning. As a matter of fact, it has been shown in Sect. 7.6 (see (7.11, 13)) that the drift velocity of a Rossby soliton – as opposed to that of a linear Rossby wave of finite length – does not depend on the average velocity of the flow. This has been confirmed by a numerical calculation of the drift of a planetary Rossby vortex having the parameters of the JGRS [10.16–18].

The other large and long-lived Rossby anticyclones in the atmospheres of Jupiter and Saturn, as shown in Sect. 2.2, have much in common with the JGRS. Thus the above solitonic model can be applied to them, too, including the Great White Ovals of Jupiter, the JLRS vortex, the Brown Spot of Saturn and some others. Their lifetimes, as shown in Tables 2.1, 2, are much greater than the lifetime of a linear Rossby wave packet, so one may call them long-lived.

In the light of the discussion in Sect. 9.6, the long-lived Rossby vortices in the atmospheres of giant planets, like those in the laboratory, may be said to exhibit the vortex-wave dualism. On the one hand, they are genuine vortices (the linear velocity of their rotation being much higher than their drift velocity). On the other hand, the anticyclones live long because, contrary to the cyclones, their dispersion is counterbalanced by the nonlinearity (while the cyclones decay relatively quickly due to the dispersion of Rossby waves).

As regards our own planet, the situation is different. The lifetime of a linear Rossby wave packet in terrestrial atmosphere, according to Table 2.1, is about two weeks which is quite comparable to the observed lifetimes of terrestrial Rossby cyclones and anticyclones – that is, the atmospheric Rossby vortices on the Earth do not exhibit solitonic properties, probably because of the too high curvature of our planet.

The above discussion shows that the simulations of the natural vortices such as the JGRS by means of two independent approaches, laboratory and numerical, yield qualitatively the same results (within the limits of their comparability). On the other hand, the major problem concerning the uniqueness of the JGRS vortex over the planet perimeter has been solved only by the laboratory method, which demonstrates once more its advantage. In our opinion, the different value of these two approaches in answering the question of why the JGRS is unique is due to the better effectiveness of the laboratory model where, just as in the natural conditions, the initial width of zonal flow velocity profile is less than the size of the vortices to be generated; this is why flow profile width increases with the size of the created vortices, which, as we believe, gives rise to the experimentally observed switchovers between vortex chains containing different numbers of vortices and leads to the emergence of a unique vortex over the system perimeter. In the numerical calculations [10.16–18], contrary to our laboratory experiment, the width of the zonal flow profile was assumed to be greater than the size of the long-lived vortices, and independent of it. There can hardly be any doubt that a more adequate numerical model will yield the same results as the laboratory experiment. (In this connection, see also the description of the results of another numerical calculation given at the end of Sect. 9.4.)

10.2 An Alternative Model of the JGRS: Numerical Calculation

Contrary to the above solitonic model of the JGRS and other, physically similar, vortices, the author of [10.25] discusses another model which has no place for Rossby solitons since Rossby waves are "forbidden" by a "voluntary act". This model is as follows.

A two-dimensional, double-layered Jovian atmosphere is considered: the upper thin layer of clouds (where the JGRS vortex is located) rests upon the thick lower layer which contains zonal flows (the mechanism of their excitation is not discussed). It is assumed for simplicity that both layers have no curvature and are contained between two coaxial vertical cylinders, rather similar to the experiments of [10.26–29]. The treatment is based on the equation of potential vortex conservation (5.13), where the denominator is expanded into a series and the terms containing $\delta H/H_0$ in powers higher than 1 are discarded. This means neglecting the scalar (KdV) nonlinearity, which, as has been shown above, creates the asymmetry in the properties of cyclones and anticyclones. As we know from Sect. 5.5, this (quasi-geostrophic) approximation is justified only for small vortex sizes: $a < r_R$, r_i. As regards the JGRS vortex whose characteristic size is $a > r_i$,[1] this approximation is of course not very convincing.

Further, one more artificial assumption is made: that no Rossby waves can propagate in the medium. This assumption also does not correspond to the actual conditions on planets and contradicts the direct observations of Rossby waves in the atmosphere of Saturn [10.30, 31] – see Fig. 5.1.

The zonal flows introduced into this system are assumed to be stable. This means that the problem of vortex generation by the flows is not considered. Only the possibility of the coexistence of a "trial" vortex, which is mentally introduced into the system, with the zonal flows is studied. With these reservations, we shall now describe the main results of [10.25] as they are of a physical interest within the scope of our discussion. These results are as follows.

If a flow around a cylinder axis, having one region of velocity shear, is set up in the system, and if two equal vortices with opposite vorticity signs are placed at the system perimeter (at the "latitude" of the velocity shear), then the numerical calculation shows that only that vortex survives whose vorticity (at its center) has the same sign as the vorticity of the flow. (Of course, there is the very essential circumstance that the vortex and the flow have been chosen so as to fit each other well.) The vortex of the opposite polarity decays.

The major result is the fact that a trial vortex of *either* vorticity can coexist with a flow – it is only required that their vorticity signs be the same. On the other hand, the JGRS vortex has a *definite* polarity – it is an anticyclone. The lack of the cyclone-anticyclone asymmetry in the results obtained in [10.25] is a direct consequence of the problem formulation adopted there.

If, in the above situation, two vortices of the same polarity are placed at different but close enough parallels, then, drifting each at the respective local velocity of the flow (there is no β-effect in the system), they will eventually get close to each other and finally merge, as the calculations show. This result obtained in [10.25] implies an interesting consequence: if the flow contains some small-scale turbulence then vortex merging occurs in times shorter than the characteristic viscous time. In other words, the presence of such turbulence makes invalid the Larichev theorem [10.32] stating the adiabatic invariance of closed potential vorticity isolines (that is, that such lines can switch to other closed or open configurations only on the scale of viscous times). This result coincides with that considered in Sect. 7.5 when the experiments reported in [10.26, 27] were discussed.

---

[1] This pertains to the characteristic size of pressure or velocity perturbation; the visual size of the JGRS is even greater than r_R.

It should be emphasized that vortex merging under the conditions simulated in [10.25] occurs only because they are placed at different (albeit close) parallels. Were the vortices placed at the same parallel, they would drift at the same velocity and never come together to merge. Identical results would be obtained with more vortices placed at the same parallel. One could also place two vortices at one parallel and two more at another, close to the former – so that two vortex pairs would merge independently. One would then obtain two identical vortices at the same parallel, rotating stationarily about the system axis and never getting close to each other. Obviously, such results are in no way related to the problem concerning the *uniqueness* of the JGRS over planet perimeter.

Now, if a zonal flow with a radially periodical vorticity is set up in the system, and two trial perturbations of opposite vorticities are superimposed diametrically on the flow, their dimensions being small along the azimuth and large (covering the entire system) along the radius, then such perturbations will split up into a number of smaller ones. The emerging small vortices are localized in the areas of "appropriate" vorticity, while the vortices of "inappropriate" vorticity are destroyed. After some time, all the zones in the radially periodical flow become filled with vortices of appropriate signs. It should be added that in the case when more than two trial perturbations are inserted along the azimuth, a chain containing quite a few vortices at the same parallel can appear. Thus, even though the study described in [10.25] seems to be inadequate to the physical conditions actually existing on planets, its results are of an undoubtful physical interest, making one wish the work to be continued.

10.3 Solitonic Vortices in the Oceans

The longest-lived among the oceanic vortices are the intrathermoclinic lenses which are Rossby anticyclones (Sect. 2.3.3). Their lifetimes can be up to several years [10.33, 34] while the characteristic dispersion time of a linear wave packet (2.21) is only about three months – that is, at least an order of magnitude less (Table 2.1). Consequently, an anticyclonic lens is a nonspreading – that is, solitonic – Rossby vortex. Cyclones are very rare among the lenses, which additionally supports the view that lenses are Rossby solitons. Besides, the lenses are vortical solitons since they carry their water over thousands of kilometers (this is found from water analysis which helps identify the place of their birth). They drift westwards, just as Rossby vortices should, and have dimensions of $a \gtrsim r_i$.

A detailed theoretical study treating the anticyclonic oceanic lenses as Rossby solitons was carried out in [10.35]. In particular, the spatial profile of an actual lens was analyzed and found to be virtually the same as the theoretical profile of a smooth scalar Rossby soliton of the highest possible amplitude ($h_0 \approx 1$). The latter profile, as shown in [10.35], approximately coincides with (5.29, 30), notwithstanding the fact that the result obtained in [10.22, 23] and given by (5.29, 30) corresponds to the approximation $h_0 \ll 1$. The author of [10.35] further compares the theoretical profile of a limiting-amplitude soliton with the profile of a similar soliton observed experimentally [10.36] ($a > r_R$, $h_0 \approx 1$) and points out their agreement (see also [10.37]). Comparisons of height profiles $h(r)$ as well as angular rate profiles $\omega_r(r)$ for anticyclones are given in Figs. 9.16 (the laboratory) and 10.1 (the ocean). It should be also recalled that, as pointed out in Sect. 9.4, the large-amplitude Rossby soliton observed in the experiments with small initial layer thickness ($\delta H > H_0$) appears a good physical model for the anticyclonic oceanic lenses whose thickness tends to zero at their periphery.

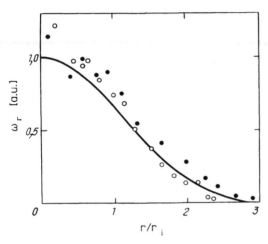

Fig. 10.1. Angular velocity profile $\omega_r(r)$ of an actual oceanic lens (*dots*) and the theoretical profile $\omega_r(r)$ of a smooth Rossby soliton having the highest possible amplitude (*solid curve*). From [10.35]

The good agreement of the theory with the simulation experiments and with the observations in the oceans serves to support the solitonic concept of the oceanic Rossby vortices. It is reasonable to suppose (in accordance with [10.35]) that, aside from the possible specific features in the generation of the lenses, the primary cause of their cyclone-anticyclone asymmetry (virtually all of them are anticyclones) is the same scalar nonlinearity which is responsible for the similar asymmetry of the long-lived vortices in the atmospheres of giant planets. For anticyclones, it compensates the dispersion and so they live long; for cyclones, there is no such compensation and thus no long-lived cyclonic lenses are observed.

It should be noted that the characteristic size of the observed lenses becomes smaller towards the north in accordance with the decreasing Rossby-Obukhov radius, $r_R \propto f^{-1}$, that is, in accordance with the theory in question.

A theory of oceanic lenses was also constructed in [10.38, 39]; it contains a more detailed treatment of the various factors entering the equations, including viscosity. These studies support the viewpoint that anticyclonic lenses are Rossby solitons and the above-mentioned reason of their cyclone-anticyclone asymmetry. Detailed computations, compared with the laboratory experiment and with actual observations, show that the lifetimes of anticyclonic lenses are determined by viscosity and are much greater than the dispersion time of a linear Rossby wave packet [10.38, 39].

It is important to draw the reader's attention to the fact that a good agreement of the theory of intrathermoclinic anticyclonic lenses with what is observed in the oceans is achieved only if the three-dimensional (baroclinic) model is adopted, with $r_R \rightarrow r_i$ and $V_R = \beta r_i^2$. The situation here perfectly resembles that discussed in Sect. 7.4 in relation to long-lived vortices of the JGRS type. (Oceanic lenses are also discussed in [10.40–44].)

There are also cyclones among the long-lived oceanic vortices. Such are the rings of the Gulf Stream (Sect. 2.3.2). Similar to lenses, their dimensions are $a \approx r_i$, and (in some cases) they have very high ratios between the linear velocities of their rotation and drift – of the order of several dozen. Although the Gulf Stream cyclones are not among the longest-lived oceanic vortices (their lifetimes are shorter than those of intrathermoclinic lenses), they nevertheless live several times longer than linear Rossby wave packets. The authors of [10.45, 46] explain this effect as follows: if the ratio of the linear velocity of vortex rotation to the velocity of its drift is very high (this ratio can be up to 25–30 in rings and lenses), the lifetime of even a cyclonic vortex of size $a \approx r_i$ rises substantially due to the vector nonlinearity. However, the effect of the vector nonlinearity is qualitatively different: it can prolong the lifetime of a ring only within an order of magnitude (Table 2.3, Sect. 2.3.2). To put it otherwise, the vector nonlinearity does not counterbalance the dispersion but only weakens its effect, so that the Gulf Stream rings decay all the same as a result of dispersion spreading. The scalar nonlinearity, contrary to its vector counterpart, *completely* counterbalances the dispersion and gives rise to anticyclonic solitons whose lifetimes are determined by viscosity rather than dispersion. (Recall that *anticyclonic* Gulf Stream rings die early due to purely geographic reasons – Sect. 2.3.2.) Thus, when applied to the Gulf Stream rings, the term "long-lived" sounds rather weak as compared to the JGRS case where even the period of its observation is thousands of times larger than the dispersion spreading interval of a linear wave packet.

The difference in the behavior of cyclones and anticyclones is also manifested in the following effect. During its dispersion decay, a cyclone radiates Rossby waves and the absolute value of its vorticity $|\text{rot}\,v|$ decreases. Since, however, the potential vorticity (5.13) is to be conserved, the radiating cyclone is displaced towards higher absolute values of the Coriolis parameter $|f|$, that is, towards the system rotation axis (from the equator to the pole), the rate of the displacement depending on the radiation intensity [10.47]. On the other hand, an anticyclone, being a Rossby soliton, radiates no waves and therefore moves only along the parallel.

As regards the open sea vortices, it was already mentioned in Sect. 2.3.1 that they usually live not very long, which is due, in particular, to their close interaction with oceanic currents whose directions are different at different depths.

Thus the rules governing the lifetimes of the major types of synoptic vortices in the oceans can be explained in a simple and consistent manner on the basis of the nonlinear dynamics discussed here.

Additional explanations for some of the material discussed in this chapter are given in Supplement 10.1.

11. Dipolar Rossby Vortices

Experiments with monopolar Rossby vortices have been described in Chaps. 7, 9. This chapter describes experiments with vortices of different structure – dipoles, consisting of a cyclone-anticyclone pair. Such structures can also be solitons, as mentioned in Sect. 5.5. (They are called solitary waves in the original theoretical study [11.1]. See Sect. 5.4 on our attitude to such terminology.) We conducted an experimental search for such structures using first our small paraboloid (in the preliminary experiments) and then the large one (Table 6.1). Technique "a" is used in both series of experiments: a perturbation extended along the parallel is generated with the pumping disk and later shapes itself into a vortex pair. We shall now describe the experiments in the above order.

11.1 Preliminary Experiments

The experiments with the small paraboloid, as a matter of fact, yield negative results [11.2]: they show that, while a system of two coupled opposite-polarity vortices can be produced (Fig. 11.1), it is not a soliton. This follows from the fact that the lifetime of the vortex pair after which its cyclone becomes dispersed (the anticyclone remaining almost intact) is much shorter than the lifetime of a monopolar (anticyclonic) Rossby soliton. This means that under the conditions of these experiments, featuring a sharp cyclone-anticyclone asymmetry, monopolar solitons are formed as a manifestation of the scalar KdV nonlinearity which prevails under (5.21) when vortex size is considerably larger than the Rossby-Obukhov radius ($a > r_R$). This kind of nonlinearity compensates only the anticyclone's dispersion, making possible for it to shape itself into a soliton. The cyclone is destroyed by this nonlinearity which, for $a > r_R$, thus prevents the other (vector) nonlinearity from manifesting itself. The latter must, in turn, predominate for $a \lesssim r_R$ and can, according to [11.1], produce a dipolar Rossby soliton.

By the physical nature of these two nonlinearities, one of which (scalar) shows a strong dependence on liquid depth and the other (vector) only a weak dependence, it is natural to expect that the result of their competition might be different if the liquid depth were increased significantly, since the value of r_R would then rise and the ratio a/r_R might become smaller. This is what we did; in addition, we switched to the large paraboloid (Table 6.1) in order to ensure

Fig. 11.1. Example of a coupled cyclone-anticyclone structure which is not a soliton and therefore decays quickly due to the scalar nonlinearity (the small paraboloid, $H_0 \approx 0.5$ cm; the characteristic size of the structure exceeds r_R). From [11.2]

that, even with increased depth, the radius r_R remained small compared to the vessel curvature radius. The experiments with the large paraboloid were carried out for a wide range of fluid depths ranging from 0.4 to 5.5 cm [11.3–5]. Their results are described in the following sections for low and high values of H_0.

11.2 Dipolar Vortex Decay for Moderate Liquid Depths

First, we are going to discuss the experiments with moderate liquid depths, $H_0 < 2.5$ cm (Fig. 11.2).

1) A simple and reliable way of producing a dipolar Rossby vortex was found to be as follows. An extended cyclonic perturbation in the parabolic rotating liquid layer is generated with the pumping disk for about 5 s. This perturbation may be viewed as a highly elongated oval, the liquid flowing in opposite directions along its long sides, which creates a counterflow-like structure. Soon after the disk is switched off the flows thus excited generate two paired (dipolar) vortices. These two dipoles move along the parallel in opposite directions, their polarizations corresponding to the relative motion of the vortex and its environment.

Namely, in the dipole that is drifting westward (that is, lagging behind the surrounding liquid), the cyclone is the outer vortex (that is, the one farther from the paraboloid center), while in the eastward-drifting dipole the outer component is the anticyclone.

2) The size of these dipoles (the distance between cyclone and anticyclone centers) is $a \lesssim (1.5 - 2)r_R$.

3) The westward drift velocity of the dipoles is larger than the Rossby velocity and equals approximately $|V_{dr}| \approx 1.5V_R$.

4) The eastward drift velocity of the dipoles varies in a wide range but usually does not exceed the Rossby velocity. Properties (3) and (4) are in a good agreement with the theoretical relationships (5.23) for a stationary dipolar soliton emitting no Rossby waves by Cherenkov radiation.

5) The vortices carry trapped particles as they travel along.

6) The paired vortices exhibit a clear cyclone-anticyclone asymmetry: the cyclone decays quickly and only the anticyclone is left from the dipole, living relatively long. (Under the conditions of Fig. 11.2, the longest-lived vortex is the anticyclone of the dipole which drifts westwards; it has a sufficiently large amplitude.) This can be explained by the predominance of the scalar nonlinearity which tends to assist dispersion in destroying cyclones instead of counterbalancing it. The dipoles are not solitons under the above experimental conditions (the liquid layer is not thick enough).

Another interesting result should be mentioned. If there is a liquid depth gradient directed to the vessel center and a westward-drifting dipolar vortex is observed, then, after the dipole has decayed, the remaining vortex (the anticyclone) slows down and begins drifting eastwards – just as a monopolar vortex should behave.

11.3 Solitonic Properties of Dipolar Vortices for Large Liquid Depths

Let us now turn to the experiments with thicker liquid layers, $H_0 \gtrsim 2.5$ cm (Figs. 11.3, 4). There is no clearly discernible cyclone-anticyclone asymmetry in this case: the lifetime of the vortex pair excited in the above manner is no more limited by cyclone decay. This can be explained by the predominance of the vector nonlinearity, which gives rise mainly to dipolar vortices. The maximum linear velocity over the vortex profile observed for the vortex in Fig. 11.4 is approximately 10 cm/s which is about a factor of 7 greater than dipole drift velocity. Under the conditions of Fig. 11.3, the vortex size is $a \approx 1.3r_R$, and the time during which the dipole exists as a vortex containing trapped particles is about 15 s. This time is limited by partner separation due to different drift velocities of cyclones and anticyclones, see Fig. 9.11.

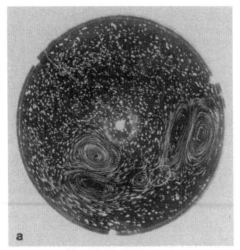

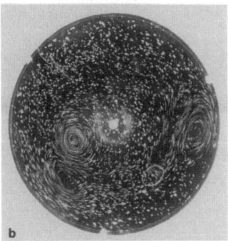

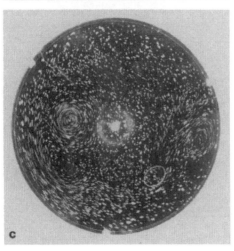

Fig. 11.2 a-c. Two dipolar Rossby vortices produced with one pumping disk (the large paraboloid, thin layer $H_0 = 1$ cm=const). The disk (bottom right of the picture) rotates in the cyclonic direction for about 5 s, creating a cyclonic perturbation extended along the parallel, which then divides in two cyclones. Next, an anticyclone is born in the vicinity of each of the two cyclones, thus forming two dipoles (the cyclones are darker than the anticyclones in the picture); the outer vortex (that which is farther from vessel center) is cyclonic in one of them (the one drifting clockwise, or westwards) and anticyclonic in the other (drifting counter-clockwise, or eastwards). The time intervals are 4 s between frames **a** and **b** and 5 s between **b** and **c**

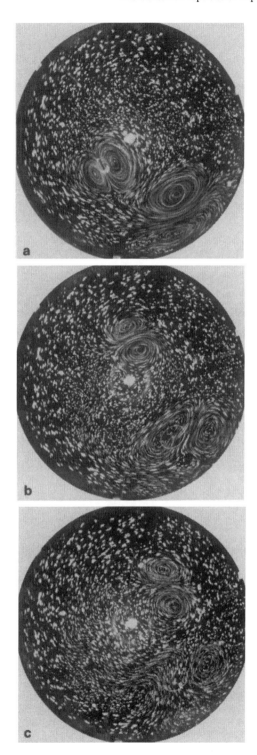

Fig. 11.3 a-c. A long-lived dipolar Rossby vortex drifting westwards, counter to vessel rotation; the partners diverge in the course of their drift (large paraboloid; $H_0 = 2.5$ cm at $\varrho = D/2\sqrt{2}$ where $\partial H_0/\partial y = 0$; the times elapsed are 8 s between pictures **a** and **b** and 7 s between **b** and **c**)

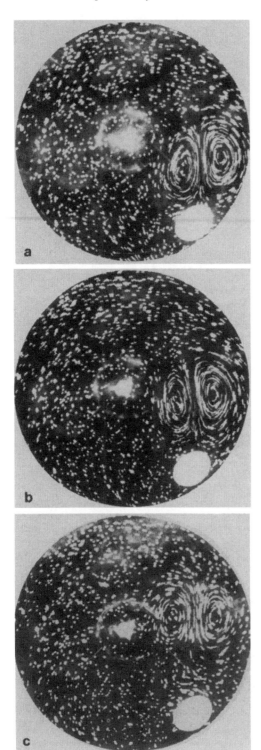

Fig. 11.4 a-c. A long-lived dipolar Rossby vortex drifting eastwards, along vessel rotation (the large paraboloid; $H_0 = 4$ cm at $\varrho = D/2\sqrt{2}$ where $\partial H_0/\partial y = 0$; the time intervals are 2 s between taking pictures **a** and **b** and 8 s between **b** and **c**)

The drift properties of dipolar Rossby vortices are illustrated in Fig. 11.5 showing the ratio of their drift velocity to the Rossby velocity as a function of layer thickness H_0. Contrary to the case of monopolar cyclones and anticyclones which drift only in the direction of linear Rossby waves, dipolar vortices can drift in either direction and, besides, their westward drift is faster – its velocity exceeds V_R – that is, they possess the properties of dipolar Rossby solitons described by (5.23).

Under conditions close to (5.22), their lifetime is appreciably greater than that of the cyclones and even approaches the lifetime of the long-lived anticyclones. So these dipoles exhibit the manifold of properties which makes them very similar to the dipolar solitons [11.1] described by the Charney-Obukhov equation (5.1). This is another example illustrating the significant influence of Rossby wave dispersion on the properties of Rossby vortices. The reciprocal influence of the vortical properties of a Rossby soliton on its structure, dynamics, and the very fact of its existence, has been discussed in Chap. 9. On the whole, these are the two sides of the vortex-wave dualism in geophysical hydrodynamics.

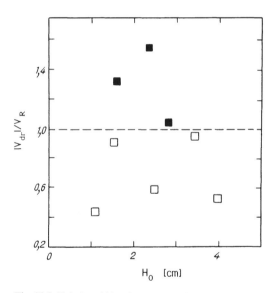

Fig. 11.5. Relative drift velocity V_{dr} of dipolar Rossby vortices as a function of unperturbed layer thickness (the large paraboloid, $\Omega = \Omega_0$) for dipoles drifting westwards (*filled squares*) and eastwards (*open squares*). Each point is a result of averaging over several runs at each value of H_0

It is interesting that the above method can produce, under certain conditions, not just two dipoles but a whole "chessboard" of vortices – Fig. 11.6a (such a vortical ensemble is called a "modon sea" in the theoretical study [11.6]). The experiments show that, for moderate layer thickness, only the anticyclones survive of all this "sea" after some time while the cyclones decay – Fig. 11.6b.

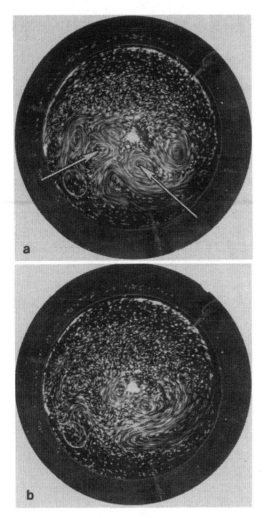

Fig. 11.6 a,b. Evolution of a "modon sea" of cyclones and anticyclones (the large paraboloid, anticyclones marked with *arrows*): the cyclones disappear rapidly and after 5 s (**b**) only two anticyclones remain of the initial six vortices (**a**)

This fact, observed in our experiments, is related to the cyclone-anticyclone asymmetry and to the collective properties of the vortices (their interactions).

The dipolar vortices described here are quite different from the "deep-water" dipolar vortices observed in the experiments of [11.7, 8] where the β-effect is very weak and the characteristic vortex size is much smaller than the Rossby-Obukhov radius, $a \approx r_R/6$ (the liquid is vertically homogeneous). According to (5.22), only the vector nonlinearity can be significant under such conditions. The lifetime of the vortices observed in [11.7, 8] is shorter than the linear Rossby wave dispersion time which is easy to evaluate using (5.16) and a simple estimate of the β-effect due to liquid surface curvature. Therefore these vortices cannot be classified as solitons. The authors of [11.7, 8] call them modons, a widely used term.

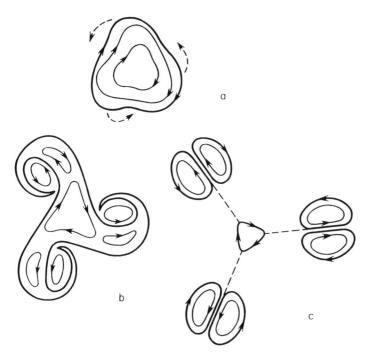

Fig. 11.7. Formation of three dipolar vortices as a result of quickly pulling out a triangular tumbler formerly immersed in the liquid (the upper picture is the initial perturbation). From [11.9]

Figure 11.7 is a schematic illustration of the generation of three dipolar vortices in a rotating liquid in the experiments of [11.9]. The rotating liquid is excited by quickly pulling out a previously immersed triangular tumbler. The subsequent liquid motion around the triangle eventually produces the pattern shown in the figure. An additional feature of such modons is their capability to move in different directions, not just along the east-west line. This is due to the weakness of the β-effect. Laboratory simulation of oceanic dipolar vortices is described in [11.10].

Fig. 11.8. Wake behind a round cylinder moving in a thin film of soap solution (the picture shows the light interference pattern determined by the perturbations of the film thickness and hence by the motion of the liquid). From *top* to *bottom*: Karman street of monopolar vortices with alternate polarity, appearing in the case of constant cylinder velocity; and streets of dipolar vortices observed in the case of varying (oscillating) velocity. From [11.11]

All the experimental results described above show that dipolar Rossby vortices undoubtedly display certain "attractor" properties: if the liquid depth is not too small, extended perturbations clearly tend to develop into dipolar rather than monopolar vortices. Therefore, in order to obtain monopolar cyclones or anticyclones, the special techniques described in Sect. 6.2 ought to be used. Otherwise (if, for example, one uses techniques "a" or "b" instead), dipolar vortices are generated, such as those obtained in our experiments illustrated by Figs. 11.2–4 or in the experiments reported in [11.7–10].

The "attractor" properties pertain to dipolar vortices in general, not only to Rossby vortices. This is confirmed both by the experiments described in [11.9, 10] and by those described in [11.11] where unstable Karman streets of vortices were investigated. Figure 11.8 shows one of the results from [11.11], the formation of dipolar vortices from the "cyclones" and "anticyclones" of a Karman chain in nonrotating liquid (here the terms "cyclones" and "anticyclones" are used only to indicate vortices of opposite polarity since the system does not rotate).

The dynamics of dipolar vortex interactions was explored by numerical calculation in [11.12]. Space observations of oceanic dipolar vortices are described in [11.13, 14]. The reader is also referred to [11.15] where, based on the theory of dipolar Rossby vortices, a theory of the spectrum of astrophysical turbulence and the observed mass spectrum in the Universe is proposed.

In conclusion, the original experiments on the observation of tripolar vortices should be mentioned [11.16, 17]. It is instructive to compare their results with the "modon sea" pattern (Fig. 11.6a) and with the tripolar vortex shown in Fig. 8 in [11.2].

12. Shallow-Water Simulation of Drift Vortices and Solitons in Magnetized Plasma

The physical analogy outlined in Chap. 5 makes it possible to simulate drift (or gradient) vortices and solitons in magnetized plasma, using simple experiments with rotating shallow water. The results of such simulations are presented in this chapter.

12.1 Prediction of Drift Soliton Properties Based on Shallow-Water Simulation. Drawbacks of the "Purely Wave" Concept

The theoretical wave concept which existed prior to our experiments included the following main principles.

1. A monopolar drift soliton is a solitary drift wave propagating independently of medium particles.

2. The transverse size of a soliton (its diameter) depends on the amplitude of its potential φ_0 and equals, according to (5.30),

$$a \approx \sqrt{3} r_\mathrm{L} (e\varphi_0/T_\mathrm{e})^{-1/2} . \tag{12.1}$$

The soliton is greatly stretched along the magnetic field ($k_z \ll k$).

3. Two solitary drift waves pass through each other freely upon collision [12.1–3], just like the classical one-dimensional KdV solitons [12.4, 5].

It was also supposed that there exists a stable dipolar drift soliton which has the form of a symmetrical pair containing opposite-polarity vortices, a cyclone and an anticyclone [12.6–9]. It is a vortical soliton, carrying trapped particles with it.

Now let us compare the above principles with the results obtained in our experimental studies of Rossby solitons physically analogous to drift solitons in magnetized plasma. Recall that these studies are carried out under the conditions $f\tau_\nu > f\tau_\mathrm{E} \approx 10^2 \gg 1$, the analog of the condition of plasma magnetization $\omega_\mathrm{B}\tau_\mathrm{i} \gg 1$ (see (4.7), (5.35) and the beginning of Sect. 5.2).

It is easy to see that the results of simulation experiments described in Chaps. 9, 11 are in a fundamental contradiction with the theoretical principles (1–3) cited above.

First of all, our experiments indicate that a monopolar Rossby soliton is a "genuine" vortex which contains trapped particles of the medium and carries them along over large distances as it drifts. It is vital for the "life" of such a soliton – and hence for a monopolar plasma drift soliton, too – that the particle capture condition (2.15) is fullfilled, i.e., that the amplitude is sufficiently large. Otherwise, the drift soliton, having a low amplitude, will quickly disperse due to the radial gradient of the drift velocity V^*. Thus the first principle of the "purely wave" concept is invalid.

The laboratory simulations show that the expected size of a drift soliton (exceeding r_L) is not related to its amplitude by (12.1). Drift solitons satisfying (12.1) are very rare exceptions and can occur in plasma with a very low probability. Therefore, the second principle of the wave concept is also invalid.

Our experiments indicate that drift soliton collisions must be inelastic, that is, the solitons must either merge upon collision or destroy each other. Thus the third principle of the wave concept is wrong, too.

Finally, our simulation shows that, for $a > r_L$, the dipolar drift solitons which correspond to the widely used Hasegawa-Mima equation [12.6–9] must degrade into monopoles. This is because only the vector nonlinearity is taken into account in the Hasegawa-Mima equation, while the scalar one which must prevail for $a > r_L$ is disregarded. The scalar nonlinearity will probably give rise to a monopolar drift soliton, similar in principle to a monopolar Rossby soliton.

The decay of dipolar vortices in a plasma was studied by numerical calculation in [12.10, 11]. It was shown that a dipolar vortex which would be a dipolar drift soliton of the Hasegawa-Mima equation if there were no scalar nonlinearity, will actually behave as shown in Fig. 12.1 when this nonlinearity is taken into account. The figure shows three consecutive patterns of a structure which initially has been a dipolar vortex. The initial characteristic size of this vortex is 4.5 times the characteristic ion Larmor radius. One can see how, under the influence of the scalar nonlinearity, one of the two vortices quickly decays, leaving a long-lived *monopolar* vortex. Under the model assumption adopted in [12.10, 11], the long-lived vortex is an anticyclone. In other circumstances, it may be a cyclone. This is discussed in detail in Supplement S5.4. The qualitative similarity between the patterns of vortex pair evolutions in Figs. 12.1 and 11.2 is obvious: a dipolar drift "soliton" in plasma, large with respect to the Larmor ion radius, decays in the same way as the corresponding dipolar Rossby vortex in shallow water. This decay, similar to the Rossby vortex case, is facilitated by divergence of the partners due to their different drift velocities (Figs. 9.11, 11.3).

Thus the simulation experiments with Rossby solitons in rotating shallow water lead to a qualitatively different view on drift solitons in magnetized plasma and allow one to predict their main properties with enough confidence. This should

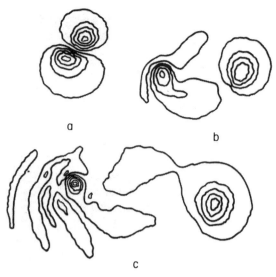

a

b

c

Fig. 12.1. A dipolar drift vortex in magnetized plasma at different stages of its decay due to a scalar nonlinearity: moments of time 0.8 (**a**), 4.8 (**b**), and 9.6 (**c**), measured in units of $(a/r_L)^2 \omega_B^{-1}$ from the initial moment when the dipole was symmetrical and its radius was $a = 4.5 r_L$; the upper vortex is anticyclonic at stage (**a**). From [12.10, 11]

greatly facilitate the investigation of drift solitons in direct plasma experiments and help in further development of the theory.

However, this is not all the benefit plasma physics can draw from shallow-water simulations. In fact, it turns out that the Rossby vortices studied in the simulations behave, as regards their mutual collisions, in a way similar to plasma convective cells whose collisions were observed in the recent experiments [12.12, 13]. A convective cell is a domain, greatly extended along the external magnetic field, where an extremum of the potential has been created by means of external boundary conditions – to put it otherwise, a domain with a non-monotonous electric field crossed with the magnetic field. The drift motion of plasma particles along the equipotentials in this domain is somewhat similar to particle motion in a vortex; the vorticity sign of the motion depends on the direction of the electric field in the cell. It is shown in [12.12, 13] that convective cells of the same sign merge – just like the Rossby solitons we have described.

Another interesting fact is that streamline reconnection which occurs during the collision (and merging) of Rossby solitons is physically analogous to the reconnection of magnetic field lines in plasma and magnetohydrodynamic configurations, as pointed out in [12.14].

It should be noted that the theory of plasma drift solitons has been revised recently [12.15–18] (the discussion of nonlinear drift waves in Chap. 5 is presented with the revisions taken into account). The following new results pertaining to the conditions for a monopolar drift soliton to exist in a magnetized plasma have been obtained.

a) Soliton amplitude must be sufficiently large:

$$(\delta n)_0/n_0 \approx e\varphi_0/T_e \gtrsim a/R \tag{12.2}$$

where R is the curvature radius of plasma density and temperature profile (see footnote 1 in Sect. 5.2). It is easily seen that the condition (12.2) means that the linear velocity of particle rotation in an electric field (with a potential φ_0) which has a characteristic length of a and is crossed with a magnetic field must be greater than the drift velocity, see (5.10). But this is equivalent to the condition (2.15) of particle capture by the vortex. Thus the monopolar drift soliton in question contains trapped particles (that is, possesses a separatrix).

b) The characteristic soliton size must satisfy the condition

$$a < (r_L^2 R)^{1/3} \tag{12.3}$$

which is analogous to the condition (5.31) imposed on Rossby solitons.

Obviously, both these statements of the revised theory are quite consistent with the predictions made on the basis of our laboratory simulation.

12.2 Vortical Mechanism for the Enhanced Diffusion of Plasma Across a Strong Magnetic Field

The results of our laboratory simulations may have an interesting application to the problem of controlled thermonuclear fusion since they can be used to predict a new vortical mechanism of fast plasma diffusion across a strong magnetic field.

Experimental investigations of the spectra of low-frequency plasma oscillations in American and French tokamaks using Raman scattering techniques indicate that the observed oscillations result from the development of the drift instability and that plasma confinement time is inversely proportional to the degree of drift turbulence [12.19–21]. The latter is illustrated by Fig. 12.2 showing that the inverse plasma confinement time in the TFR tokamak is proportional to the value of $\langle(\delta n)^2\rangle/n_0^2$ which characterizes the level of low-frequency (drift) turbulence [12.20]. Experimental evidence on the existence of coherent structures (compressions) in plasma, their sizes about five times the characteristic ion Larmor radius, is reported in [12.22].

The instability, in principle, can result in a relatively small-scale turbulence, but this is not the only possible way of its development. Another scenario for instability development is the formation of large-scale structures. This is exactly what is demonstrated by the above laboratory simulation of plasma drift solitons.

The following predictions can be made on the basis of our experimental results. First, the drift instability of plasma in magnetic traps (either closed, like tokamaks, or open, like magnetic mirrors) will generate monopolar vortical

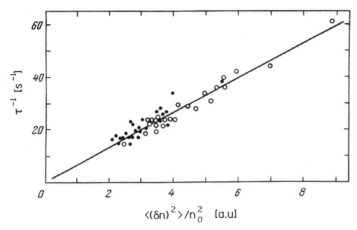

Fig. 12.2. Inverse plasma confinement time τ^{-1} as a function of the relative amplitude of low-frequency (drift) oscillations $\langle(\delta n)^2\rangle/n_0^2$, in the cases of Joulean heating (*filled circles*) and heating with an injected neutral beam (*open circles*) in the TFR tokamak. From [12.20]

solitons which are large compared to the characteristic ion Larmor radius and carry trapped particles, as well as dipolar vortices degenerating into monopolar solitons. Second, trapped particles will transfer from one vortex to another during the transition processes of streamline reconnection in the collisions (and merging) of vortical solitons, and also in the stationary regime on the borders between neighboring solitons, so that the particles will be carried over large distances across the strong magnetic field, Fig. 12.3. Such processes can therefore give rise to a very large increase in transverse diffusion and heat transfer in the plasma [12.23].

Since the classical coefficient of plasma diffusion across a magnetic field is $D_c = r_{Le}^2 \tau_{ei}^{-1}$ where r_{Le} is the electron Larmor radius and τ_{ei} is the characteristic time of electron-ion collisions, while the coefficient of the postulated vortical diffusion in magnetized plasma is $D_v \approx a^2 \tau_v^{-1}$ where the vortex radius satisfies $a > r_L \gg r_{Le}$ and τ_v is the characteristic time of interaction between vortices, it appears quite likely that $D_v \gg D_c$.

It is interesting to note that, judging from [12.21, 22], the criterion (12.2) is satisfied in the American CALTECH tokamak for the observed drift turbulence, meaning that drift solitons are quite likely to exist there. In other tokamaks [12.19, 20], the condition (12.2) is near its validity margin and thus the problem of whether drift solitons can exist there is also on the agenda.

The drift turbulence and its influence on the processes of plasma transfer across a strong magnetic field are investigated in the theoretical works [12.24, 25] on the basis of a "gas of drift solitons", similar to the above Rossby solitons. The dynamics of drift vortices (solitons) and its relation to anomalous diffusion of magnetized plasma are discussed in a recent theoretical study published in [12.26].

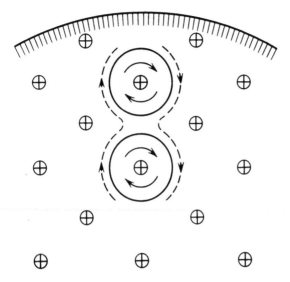

Fig. 12.3. Plasma particle transfer across a magnetic field which is directed away from the reader perpendicular to the plane of the drawing, as a result of interaction between drift vortices (their merging)

We have discussed the implications of shallow-water experiments to plasma physics in the case of free-traveling drift vortices similar to Rossby vortices. But let us now recall the experiments with vortices generated by unstable flows (Chaps. 7, 8). In accordance with what has been said above, there are reasons to believe that quite similar structures consisting of drift vortices (solitons), as well as spiral-vortex structures, will be generated in the appropriate plasma flows. If an external magnetic field is crossed with a nonuniform electric field (that is, directed perpendicular to it), a plasma drift motion with a velocity shear will arise; if the electric field has a spatially non-monotonous potential, plasma counterflows will be produced. Nonuniform electric fields are quite common in high-temperature plasma.

Another hydrodynamic model of plasma counterflows in crossed fields is the system where the liquid is rotated in the gravity field crossed with a spatially non-monotonous field of temperature gradient; this system also generates vortex chains (see [12.27, 28] and the references cited therein).

Sometimes (under certain conditions), spiral structures are observed in plasma. Comparing them with the shallow-water spirals (Chap. 8), one should keep in mind the following important fact. The spiral arms (Figs. 8.1, 2), in contrast to the vortices located between them (Fig. 8.4), are *waves* which carry no particles of the medium along. Herein lies their difference from the spiral arms, observed in some plasma experiments [12.29 – 33], which are not waves but rather the result of a radial motion of charged particles in the case where a longitudinal magnetic field is crossed by a perturbed azimuthal electric field, Fig. 12.4 and [12.29 – 33].

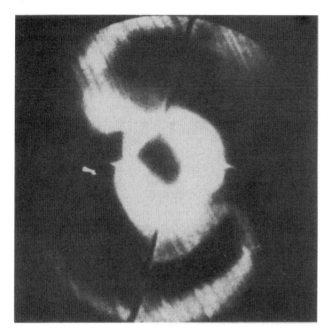

Fig. 12.4. A spiral structure in plasma. From [12.29]

13. Conclusion

We have·seen that Nature displays a wide variety of such nonlinear self-organizing coherent structures whose qualitative and quantitative study is greatly facilitated by physical simulation in rotating shallow water. The structures in question are Rossby vortices in oceans and planetary atmospheres, spiral structures in galaxies, and drift solitons and vortices in plasma.

The generation of such structures occurs on the background of flows with velocity shear and appears to result from a common mechanism – the hydrodynamic instability of differentially rotating shallow water with a free surface. Their existence is vivid evidence that instability does not necessarily lead to turbulence and chaos but can also produce large-scale structures whose characteristic sizes range from the tiny ion Larmor radius in plasma to many thousands of light-years in galactic spirals.

Coherent structures and the nonlinear processes of their self-organization have recently become an object of intense study by many researchers in physics, chemistry, biology and other areas (see, e.g., [13.1–8]). Self-organizing structures are often referred to as "autostructures", understood as localized spatial structures, stable in dissipative and non-equilibrium environments and independent (within certain limits) of boundary and initial conditions [13.8, 9]. Autostructures maintained in their stationary state by pumping from (zonal) flows may have the same properties and the same configuration as free structures which exist without pumping. The examples of such self-organizing and self-sustained structures are Rossby autosolitons, the Great Red Spot of Jupiter among them. A review of self-organization phenomena in continuous media and, in particular, under the conditions of geostrophic turbulence of Rossby waves or plasma drift waves, can be found in [13.10].

Rossby vortices are also associated with other phenomena which could not have been discussed here due to the lack of space – for example, the Rossby waves appear to play an important role in the hydromagnetic dynamo effect in astrophysical media [13.11–13].

It should be noted that coherent (vortical) structures in rotating liquids, generally different from Rossby vortices, are currently under investigation in a number of experiments which yield quite interesting results, too. One example is the experiments described in [13.14] where three-dimentional vortical structures resembling the Hashimoto soliton were observed in rotating deep water, in a regime quite different from the Rossby regime (Ro = 3 to 33). The latter soliton is a vortical string extended along the system rotation axis, with an envelope soliton in the form of a solitary loop propagating along it, string thickness being small compared to the Rossby-Obukhov radius and to the thickness of the liquid layer.

Worth mentioning is the fact that one concept in the general theory of hydrodynamic turbulence, currently in rapid development, is based on the idea that turbulence is a dynamic interaction between coherent structures and this interaction can become chaotic under certain conditions (see, for instance, the review given in [13.15]). Thus the coherent structure types described in this book may prove to be of interest for that theory as possible "elementary cells" of turbulence.

We have discussed soliton physics only as far as was necessary for the identification of the atmospheric, oceanic, and plasma vortices simulated in the experiments. An extensive body of literature is available today in this area. Among the "solitonic" natural phenomena not covered in this book are solitary tsunami waves [13.16] as well as internal-wave solitons in the atmosphere [13.17] and oceans [13.18] of our planet. These impressive phenomena are very interesting but they are relatively small-scaled, not complying with the criterion given by (2.3), so they are not affected by planet rotation and therefore fall beyond this book's scope.

The potential of the method we have used is not limited to the examples of successful hydrodynamic simulation of natural phenomena described here. The following goals for future work might be indicated, among others:

a) constructing a three-dimensional laboratory model of natural Rossby vortices based on a synthesis of the shallow water and thermogyroconvection models;

b) simulating the evolution of quasi-two-dimensional turbulence – including the formation of zonal flows and large-scale vortices;

c) simulating the processes of vortical diffusion of plasma across a strong magnetic field;

d) simulating the complicated processes of magnetic field line reconnections in the dynamics of astrophysical and laboratory plasma.

A very encouraging fact is the great heuristic importance of this method. It provides a new and deeper insight into the phenomena in question, making it possible to predict substantially new effects which can be verified through direct observation, and thus stimulates the work of both observing astronomers and physicists involved in theoretical or experimental studies.

It is a pleasure for us to finish this book with a recent outstanding result obtained with the Voyager 2 spacecraft, which is directly relevant to the book's content. This spacecraft, launched in 1977, investigated Jupiter (1979), Saturn (1982), Uranus (1986), and, before saying goodbye to the Solar system, photographed the atmosphere of Neptune in summer 1989. A giant vortex was discovered in Neptune's southern hemisphere (by the way, at 22°S, just as the JGRS), its size relative to the radius of the planet being about the same as that of the JGRS. This vortex (Fig. 13.1) was named the "Neptune Great Dark Spot"

(NGDS) [13.19, 20]. It is interesting to compare its properties with the concepts put forward in this book.

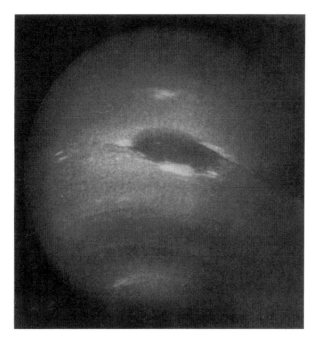

Fig. 13.1. Neptunian vortex Great Dark Spot as seen from the south. From [13.19]

First of all, it should be pointed out that, within the framework of the isothermal upper atmosphere model we use, the Rossby-Obukhov radius $r_R = (T/M)^{1/2}/f$ is the approximately same for Neptune as for Jupiter. Indeed, the atmospheric temperature is practically the same on both planets and the difference in the molecular weight of the main gas component (helium on Neptune and hydrogen on Jupiter) is compensated by Neptune's slower (by a factor of about 1.5) proper rotation. As a result, the above formula yields $r_R \approx 6,000$ km in both cases. The radius of the NGDS is close to this value. Thus, the condition $a \approx r_R \gg r_i$ is satisfied for the NGDS vortex which therefore belongs to large-scale Rossby vortices.

The next question is about the direction of NGDS rotation. There was no information on this in the first reports about the discovery of the vortex [13.19, 20]. At that time we made a confident prediction, based on the material of this book, that such a vortex as the NGDS which clearly dominates in the planet's atmosphere must be an anticyclone [13.21–23]. This prediction was subsequently confirmed by Ingersoll [13.24].

The vorticity sign of the NGDS can be inferred from the characteristic image details in Fig. 13.1. Indeed, the striped image structure is apparently evidence that there are zonal flows on Neptune and atmospheric inhomogeneities extend along them, forming the stripes over the planet perimeter just as in the cases of Jupiter and Saturn. Such a stripe is also discernible at the NGDS latitude (see Fig. 13.1 and the schematic in Fig. 13.2 where it is shown by triple dashed line). This stripe bends near the NGDS, turning southwards from its "native" parallel (down in the figure) in the western part of the Spot and northwards in its eastern part. The underlying cause of these bends can only be the existence of meridional velocity components in these areas of the NGDS, whose direction (shown by vertical arrows in Fig. 13.2) corresponds to an anticyclonic motion within the vortex which is therefore an anticyclone. Since it would seem very exotic if it existed on the background of an opposite (cyclonic) shear flow, it is natural to presume that the zonal flow around the NGDS anticyclone also has an anticyclonic vorticity sign. This assumption does not contradict the observed picture which generally resembles the situation where anticyclonic Kelvin "cat's eyes" are formed in an unstable flow of anticyclonic vorticity.

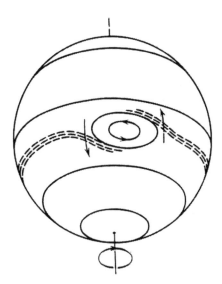

Fig. 13.2. Schematic picture of the NGDS vortex

Thus the observation of a large vortex in Neptunian atmosphere seems to confirm the idea of cyclone-anticyclone asymmetry of large-scale Rossby vortices, pursued throughout this book.

S. Supplements

The supplements include information obtained by the authors after the main body of the book had been finished as well as additional explanations pertaining to some important points.

The supplements refer to specific chapters of the book. They are numbered according to the numbering of the respective chapters. For instance, the four supplements to Chap. 5 are denoted S5.1–S5.4, etc.

S2.1 Explanation of the Euler Equation in the Form (2.1)

To understand how the Euler equation for perturbed motion is obtained in the form (2.1), let us consider a two-dimensional shallow liquid with a free surface, rotating stationarily about the η axis at an angular rate Ω_0. Denote by $\eta(\varrho)$ the equation of the intersection between the free surface and a plane passing through the rotation axis; $x = \varrho\varphi$ is the coordinate along the "parallel" on the liquid surface, φ is the azimuthal angle ("longitude"), ϱ is the distance from the rotation axis. In the frame rotating with the liquid, the Euler equation in terms of the velocity components (u, v) along the x and ϱ axes is

$$u_t + u_x u + u_\varrho v - fv = -g\eta_x \ ,$$

$$v_t + v_x u + v_\varrho v + fu = -g\left(\eta_0 - \frac{\Omega_0^2 \varrho^2}{2g}\right)_\varrho - g(\delta\eta)_\varrho$$

where the subscripts denote partial derivatives with respect to t, x, ϱ; $\eta = \eta_0 + \delta\eta$; η_0 is the unperturbed coordinate of a surface point in its stationary state; $\delta\eta$ is a perturbation of the free surface which causes perturbed motion whose velocity is (u, v). In the second equation, the centrifugal acceleration $\Omega_0^2 \varrho$ due to the system rotating like a solid body has been transferred into the argument of the full pressure gradient. We see that the free surface of a stationarily rotating liquid has the form of a paraboloid:

$$\eta_0 = \frac{\Omega_0^2 \varrho^2}{2g} \ . \tag{S2.1}$$

From here on, in order to exclude the centrifugal force due to liquid rotation, we shall consider *perturbed* motion along the surface of the paraboloid given by (S2.1), and measure the shallow water thickness perturbation δH along the normal z to that surface. Evidently,

$$\delta H = \delta \eta \cdot \cos \alpha$$

where α is the angle between the z and η axes (the polar angle complementary to "latitude"), and

$$g \, \delta \eta = g^* \delta H$$

where

$$g^* = g / \cos \alpha$$

is the joint acceleration of the gravitational and centrifugal forces. Consider the frame where the y coordinate is measured along the paraboloid generatrix (that is, along the "meridian") and the x coordinate along the "parallel" as before. The Euler equation for perturbed motion has the form of (2.1) in this frame, i.e.

$$u_t + u_x u + u_y v - fv = -g^*(\delta H)_x \,, \tag{S2.2}$$

$$v_t + v_x u + v_y v + fu = -g^*(\delta H)_y \tag{S2.3}$$

where the thickness of the "shallow water" layer is measured along the local vertical z.

S5.1 On Derivation and Analysis
of the Geostrophic Vortex Equation (5.2)

To obtain (5.2), we augment the Euler equations (S2.2, 3) with the continuity equation

$$H_t + (Hu)_x + (Hv)_y = 0 \,,$$

or

$$H_t + H_x u + H_y v + H(u_x + v_y) = 0 \,.$$

This implies

$$u_x + v_y = -\frac{1}{H}\frac{dH}{dt} \tag{S5.1}$$

where

$$\frac{d}{dt} = \frac{\partial}{\partial t} + u\frac{\partial}{\partial x} + v\frac{\partial}{\partial y} \tag{S5.2}$$

is the complete derivative.

Differentiating (S2.2) and (S2.3) with respect to y and x, respectively, and subtracting the first equation from the second, we obtain:

$$(v_x - u_y)_t + (v_x - u_y)_x u + (v_x - u_y)_y v + (v_x - u_y)u_x$$
$$+ (v_x - u_y)v_y + f(u_x + v_y) + \beta v = 0 \tag{S5.3}$$

where the relationship $f = f_0 + \beta y$ (β-plane approximation) has been used.

Let us rewrite (S5.3) in terms of the velocity curl, or vorticity. By definition, the projection of this vector onto the z axis is

$$\omega' = v_x - u_y . \tag{S5.4}$$

We then have

$$\omega'_t + \omega'_x u + \omega'_y v + \omega'(u_x + v_y) + f(u_x + v_y) + \beta v = 0 .$$

Noticing that the sum of the first three terms is $d\omega'/dt$ and that $\beta v = df/dt$, and substituting $(u_x + v_y)$ from (S5.1), we obtain

$$\frac{d}{dt} \frac{(\omega' + f)}{H} = 0 . \tag{S5.5}$$

This is the Ertel theorem (5.13), written for projections onto the z axis.

Consider a shallow water layer which in its stationary state has a uniform thickness H_0. Assume that a perturbation of the free surface $\delta H(x, y, t)$ occurs, so that

$$H = H_0 + \delta H(x, y, t) . \tag{S5.6}$$

According to (S5.5, 6),

$$\frac{d}{dt} \left(\frac{\omega' + f_0 + \beta y}{H_0 + \delta H} \right) = 0 . \tag{S5.7}$$

This means that

$$\frac{d\omega'}{dt} + h\frac{d\omega'}{dt} + \beta v + h\beta v - \omega'\frac{dh}{dt} - f_0\frac{dh}{dt} - \beta y\frac{dh}{dt} = 0 \tag{S5.8}$$

where $h = \delta H/H_0$.

Now consider the geostrophic approximation (the Rossby regime). In this regime, according to (2.4),

$$u = -\frac{g^*}{f_0}(\delta H)_y, \quad v = \frac{g^*}{f_0}(\delta H)_x , \tag{S5.9}$$

and, consequently,

$$\omega' \equiv (v_x - u_y) = \frac{g^*}{f_0}\Delta(\delta H) . \tag{S5.10}$$

Within this approximation, according to (S5.2, 9),

$$\frac{dh}{dt} = h_t .$$

Expressing the full derivative of ω' in terms of partials and making use of (S5.4, 9, 10), we get

$$\frac{d}{dt}\omega' = \frac{g^* H_0}{f_0} \Delta h_t + \frac{g^{*2} H_0^2}{f_0^2} J(h, \Delta h) \tag{S5.11}$$

where

$$J(h, \Delta h) = h_x \Delta h_y - h_y \Delta h_x . \tag{S5.12}$$

Now, with the aid of (S5.11, 9), we obtain from (S5.8)

$$(\Delta h - \frac{h}{r_R^2})_t + \beta h_x + \beta h h_x + \frac{g^* H_0}{f_0} J(h, \Delta h) = 0 \tag{S5.13}$$

where

$$r_R = \frac{(g^* H_0)^{1/2}}{f_0} .$$

and the following simplifying assumptions have been made:

$$-\Delta h h_t \ll h_t/r_R^2 ,$$
$$h\Delta h_t \ll \Delta h_t ,$$
$$-\beta y h_t \ll f_0 h_t , \tag{S5.14}$$
$$\frac{g^* H_0}{f_0} J(h, \Delta h) \cdot h \ll \beta h h_x .$$

These simplifications are justified because in our treatment we assume the following conditions to be valid:

$$h \ll 1 , \quad r_R < a , \quad \beta y \ll f_0 , \quad y < a \ll R \approx f_0/\beta .$$

The equation (S5.13) we obtained above is known as the generalized geostrophic equation. Introducing the dimensionless variables

$$x \to r_R x , \quad y \to r_R y , \quad f_0 t \to t , \quad V_R \to \sqrt{g^* H_0} \cdot V_R , \tag{S5.15}$$

the equation (5.2) we have used in Chap. 5 is finally obtained from (S5.13):

$$(\Delta h - h)_t + V_R h_x + V_R h h_x + J(h, \Delta h) = 0 . \tag{S5.16}$$

It is quite important to note that this equation is not symmetrical with respect to the replacement of cyclones with anticyclones. Indeed, this replacement leads to $h \to -h$ and $y \to -y$, since the "right-hand" triplet of the coordinate axes

x, y, z is conserved. All the terms in (S5.16) then change their signs except for the term $V_R h h_x$ which retains its former sign. The last circumstance is the underlying cause of the cyclone-anticyclone asymmetry discussed in the main body of the book. However, one might ask whether the cyclone-anticyclone asymmetry would remain in the absense of the β-effect (when $V_R = 0$). According to the above explanation, it would not. But this explanation is not sufficient. Indeed, as can be inferred from the last condition in (S5.14), in the absense of the β-effect the cubic nonlinearity term

$$\frac{g^* H_0}{f_0} J(h, \Delta h) \cdot h$$

has to be retained in (S5.13). This term is what keeps the cyclone-anticyclone asymmetry even when $\beta = 0$. And it may be far from small, since the value of h must exceed a certain threshold ($h \gg a/R$) in order to ensure particle capture by the vortex (Chap. 5). See more about the importance of taking this cubic nonlinearity into account in [S5.1].

With the aid of (S5.16), the character of the cyclone-anticyclone asymmetry caused by the β-effect can be estimated. Indeed, since the sum of the second and third terms of (S5.16) is

$$V_R h_x (1 + h) , \tag{S5.17}$$

then a *soliton*, that is, a structure which is moving at a higher speed than V_R and therefore does not radiate linear waves, is possible only for $h > 0$. To put it otherwise, only an anticyclone can be a soliton.

Let us now evaluate the ratio of the scalar nonlinearity, which is due to the term containing β in (S5.13), to the vector nonlinearity connected with the Jacobian. First of all, note that in the case when the structure in question is axially symmetrical, the Jacobian (S5.12) is known to be zero. Now, if the structure deviates slightly from axial symmetry, the Jacobian is approximately equal to any one of its components – for example, $J_1 \propto f_0 r_R^2 h^2 / a^4$ – multiplied by a/R where a is the structure size and R is the system "curvature radius", that is, the distance over which the Jacobian changes appreciably. Taking this into account, the following estimate is obtained for a nearly axially symmetrical vortex:

$$\frac{\text{scalar nonlinearity}}{\text{vector nonlinearity}} \approx \left(\frac{\beta h^2}{a}\right) \Big/ \left(\frac{f_0 r_R^2 h^2}{a^4} \frac{a}{R}\right) \approx \frac{a^2}{r_R^2} .$$

And, if the Froude number

$$F = \frac{a^2}{r_R^2} > 1 ,$$

then the scalar nonlinearity is stronger than the vector one. If, on the other hand, the structure is highly anisotropic (as, for example, in the case of a dipolar vortex), that is, if vortex geometry is far from axial symmetry, then

$$\frac{\text{scalar nonlinearity}}{\text{vector nonlinearity}} \approx \frac{a^3}{r_R^2 R} < 1 \;,$$

because, as shown in Chap. 5, in the intermediate geostrophic regime $a^3 \lesssim r_R^2 R$. This means that the vector nonlinearity is responsible for the generation of dipolar vortices. Naturally, it also controls such an axially asymmetrical process as the merging of vortices.

S5.2 Geostrophic Vortex Equation for Shallow Water in a Vessel with Inclined Bottom

In Supplement 5.1 we assumed that $H = \text{const}$ in the equilibrium state. Now suppose that the equilibrium layer thickness varies along the meridian:

$$H = H_0 + \beta_2 y \;;$$

besides,

$$f = f_0 + \beta_1 y$$

where we denote by β_1 the parameter which was formerly called β. Proceeding in the same way as in the previous supplement, we obtain

$$\begin{aligned}
\left(\Delta h - \frac{h}{r_R^2}\right)_t &+ \left(\beta_1 - \frac{\beta_2 f_0}{H_0}\right) h_x + \beta_1 h h_x - \frac{\beta_2 g^*}{f_0} \Delta h h_x \\
&+ \frac{g^* H_0}{f_0} J(h, \Delta h) = 0 \;,
\end{aligned} \tag{S5.18}$$

or, in terms of the dimensionless variables given in (S5.15),

$$(\Delta h - h)_t + (V_{R_1} - V_{R_2})h_x + V_{R_1} h h_x - V_{R_2} \Delta h h_x + J(h, \Delta h) = 0 \tag{S5.19}$$

where

$$V_{R_1} = \beta_1 r_R^2, \qquad V_{R_2} = \beta_2 \frac{f_0}{H_0} r_R^2 \;.$$

In the derivation of (S5.19), we have omitted the term $-\Delta h \cdot h_t$, similarly to the derivation of (S5.13).

The equation we have obtained differs from (S5.16) by two terms. The first of them influences the β-effect. With that term included, vortex drift velocity becomes

$$\beta = \beta_1 - \beta_2 \frac{f_0}{H_0} = H_0 \frac{\partial}{\partial y}\left(\frac{f}{H}\right) \;. \tag{S5.20}$$

This is the expression we used in Chap. 5. It is utilized in some experiments where vessels with inclined bottom ($H_0 \neq \text{const}$) are employed for creating the β-effect.

The second – new – term in (S5.19) shows the influence of β_2 on the nonlinear properties of the vortices. Like the nonlinear term related to β_1, it violates the symmetry of (S5.19) with respect to cyclone-anticyclone replacement.

The presence of a thickness gradient of the shallow water layer makes a new nonlinear effect feasible. In order to see this, consider the case when the gradients of f and H_0 are of the same sign – that is, the thickness of shallow water decreases towards the periphery (Fig. 6.1). Assume that, of the two physical factors giving rise to the β-effect, the one connected with layer thickness gradient is prevailing, that is,

$$\beta = \frac{V_{R_1} - V_{R_2}}{r_R^2} < 0 \ . \tag{S5.21}$$

Let us write the sum of the second and third terms of (S5.19) as

$$V_{R_1} h_x \left(\frac{V_{R_1} - V_{R_2}}{V_{R_1}} + h \right) \ . \tag{S5.22}$$

We see that in our present case when the first term in the parentheses is negative, the nonlinear addition will increase the absolute value of wave phase velocity only if $h < 0$, implying that only a *cyclone* can be a soliton in this case. It is seen from (S5.19) that the soliton propagates eastwards, in the direction of system rotation.

The effect discussed above has an interesting counterpart in magnetized plasma whose density (and temperature) falls off towards the periphery, just as the thickness of shallow water in our case. Under such conditions, a *cyclonic* drift soliton is feasible in a plasma, too; it is similar to the cyclone in inclined-bottom shallow water we have just discussed (see Suppl. 5.4 on this topic). This is one of the manifestations of the physical analogy between Rossby vortices and drift vortices in a plasma, deserving an experimental validation.

S5.3 The "Solid Lid" Case

The differential equation for a geostrophic vortex can be written in terms of the stream function ψ if one recalls that

$$u = -\frac{\partial \psi}{\partial y}, \qquad v = \frac{\partial \psi}{\partial x} \ . \tag{S5.23}$$

For a free-surface "shallow water", (S5.23) implies, according to (S5.9),

$$\psi = \frac{g^* H_0}{f_0} h \ .$$

Therefore one can make the replacement

$$h \to \frac{f_0}{g^* H_0} \psi \ .$$

Equations (S5.13) and (S5.18) will then look more universal and be applicable also in the case when the liquid has no free surface. Such a liquid is incompressible not only in the three-dimensional but also in the two-dimensional sense, meaning $r_R \to \infty$ and $h/r_R^2 \to 0$. The $\beta_1 h h_x$ nonlinearity also vanishes since it is related to an elevation of the free surface. A theoretical analysis shows that monopolar Rossby solitons cannot exist in this case (G. Chernikov, private communication).

S5.4 Derivation of the Differential Equation for a Drift Vortex in Magnetized Plasma

We shall consider particle motion in quasi-neutral plasma within the framework of the electrostatic approximation characterized by a two-dimensional potential $\varphi(x, y, t)$ with respect to the surrounding space. We assume the plasma to be symmetrical about the z axis which is directed along the magnetic field $\boldsymbol{B}$, plasma density n decreasing monotonously along the radius (in the y direction) while staying constant over the azimuth (in the x direction). If the magnetic field is directed downwards from the page plane, away from the reader, the ions are rotating counter-clockwise and the electrons clockwise. The x axis is directed counter-clockwise, the three axes x, y, z constituting a right-handed vector triplet (Fig. S5.1). It is known (see, for instance, [S5.2]) that drift waves can exist in such a system, propagating clockwise (in the "electron direction") around the plasma column axis.

Suppose that a plasma density perturbation occurs in the above system (Fig. S5.1), giving rise to a drift wave. The perturbed collective particle motions in the x, y plane will be assumed to be sufficiently slow so that the electrons, moving rapidly along the magnetic field, have ample time to settle to the Boltzmann density distribution

$$n(y) = n_0(y) \exp\left(\frac{e\varphi}{T}\right) \tag{S5.24}$$

where $T = T_e$ is the electron temperature (plasma ions are supposed to be "cold").

The equation of motion for the ion component of the plasma in the magnetic field, under a perturbed potential φ, is

$$\boldsymbol{v}_t + (\boldsymbol{v} \nabla)\boldsymbol{v} = \left[\boldsymbol{v}\,\omega_B\right] - \frac{e}{M}\nabla\varphi \tag{S5.25}$$

where

$$\omega_B = \frac{qB}{Mc} \ ,$$

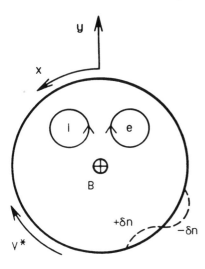

Fig. S5.1. The *circles* i and e show the rotation directions of ions and electrons in the magnetic field, V^* is the drift velocity, $\pm\delta n$ is the perturbation of plasma density

M is ion mass, q is its charge, c is the speed of light, $\boldsymbol{v}(u, v)$ is the ion velocity vector whose projections are u (on the x axis) and v (on the y axis).[1]

According to the continuity equation,

$$\frac{\partial n}{\partial t} + \text{div}\,(n\boldsymbol{v}) = 0 \ . \tag{S5.26}$$

The equations for velocity projections are, from (S5.25),

$$u_t + u_x u + u_y v = v\omega_B - \frac{q}{M}\varphi_x \ , \tag{S5.27}$$

$$v_t + v_x u + v_y v = -u\omega_B - \frac{q}{M}\varphi_y \tag{S5.28}$$

where $\omega_B = \boldsymbol{e}_z \cdot \omega_B$ and $\boldsymbol{e}_z$ is the unit vector of the z axis. Now let us differentiate (S5.27) with respect to y, (S5.28) with respect to x, and subtract the former from the latter:

$$(v_x - u_y)_t + (v_x - u_y)_x u + (v_x - u_y)_y v + (v_x - u_y)(u_x + v_y) + \omega_B(u_x + v_y) = 0. \tag{S5.29}$$

We now introduce the *velocity curl* vector rot $\boldsymbol{v}$. By definition, its projection on the z axis is

$$\omega' \equiv v_x - u_y \ . \tag{S5.30}$$

From the last equation we then obtain

$$\omega'_t + \omega'_x u + \omega'_y v + \omega'(u_x + v_y) + \omega_B(u_x + v_y) = 0 \ . \tag{S5.31}$$

---

[1] It is to be borne in mind that the vector ω_B is directed along the magnetic field (the z axis), i.e., into the opposite direction as the angular velocity vector of the ion rotation (Fig. S5.1).

In order to find $(u_x + v_y)$ we make use of (S5.26):

$$n_t + n_x u + n u_x + n_y v + n v_y = 0 \;,$$

whence

$$u_x + v_y = -\frac{1}{n}\frac{dn}{dt} \;. \tag{S5.32}$$

It is easily obtained from (S5.29) and (S5.32) that

$$\frac{d}{dt}\left(\frac{\omega' + \omega_B}{n}\right) = 0 \;.$$

We have arrived at an equation of potential vorticity conservation, similar to the Ertel equation (S5.5) in hydrodynamics. With (S5.24) taken into account, this equation becomes

$$\frac{d}{dt}\left[\frac{\omega' + \omega_B}{n_0 \exp(q\varphi/T)}\right] = 0 \;. \tag{S5.33}$$

Now we consider the differentiation of (S5.33). Note that

$$\frac{d}{dt}(\omega' + \omega_B) = \frac{d\omega'}{dt} = \omega'_t + \omega'_x u + \omega'_y v \;. \tag{S5.34}$$

Let us introduce the drift approximation, analogous to the Rossby regime, assuming the terms in the left-hand sides of (S5.27, 28) to be small compared to those in the right-hand sides – similar to the geostrophic approximation (S5.9). Then

$$u = -\frac{q}{M\omega_B}\varphi_y \;, \qquad v = \frac{q}{M\omega_B}\varphi_x \;, \tag{S5.35}$$

and

$$\omega' = \frac{q}{M\omega_B}\Delta\varphi \;. \tag{S5.36}$$

From (S.5.34–36) we get

$$\frac{d}{dt}(\omega' + \omega_B) = \frac{q}{M\omega_B}\Delta\varphi_t + \frac{q^2}{M^2\omega_B^2}J(\varphi, \Delta\varphi) \tag{S5.37}$$

where

$$J(\varphi, \Delta\varphi) = \varphi_x \Delta\varphi_y - \varphi_y \Delta\varphi_x \;.$$

Let us proceed with differentiating (S5.33).

$$\frac{d}{dt}\left[n_0 \exp\left(\frac{q\varphi}{T}\right)\right] = \left[n_0 \exp\left(\frac{q\varphi}{T}\right)\right]_t + u\left[n_0 \exp\left(\frac{q\varphi}{T}\right)\right]_x + v\left[n_0 \exp\left(\frac{q\varphi}{T}\right)\right]_y$$

After some rather simple calculations we obtain, using (S5.35),

$$\frac{d}{dt}\left[n_0 \exp\left(\frac{q\varphi}{T}\right)\right] = \exp\left(\frac{q\varphi}{T}\right)\left[n_0 \frac{q}{T}\varphi_t + \frac{q}{M\omega_B}\frac{\partial n_0}{\partial y}\varphi_x \right.$$
$$\left. - \frac{q^2 n_0}{M\omega_B T^2}\frac{\partial T}{\partial y}\varphi\varphi_x\right] . \tag{S5.38}$$

Next, (S5.33, 37, 38) yield

$$(\Delta\varphi - \frac{\varphi}{r_L^2})_t - \frac{\omega_B}{n_0}\frac{\partial n_0}{\partial y}\varphi_x + \frac{q\omega_B}{T^2}\frac{\partial T}{\partial y}\varphi\varphi_x$$
$$- \frac{q}{M\omega_B n_0}\frac{\partial n_0}{\partial y}\Delta\varphi\varphi_x + \frac{q}{M\omega_B}J(\varphi, \Delta\varphi) = 0 , \tag{S5.39}$$

where the third-order term

$$\frac{q^2}{M\omega_B T^2}\frac{\partial T}{\partial y}\varphi\varphi_x\Delta\varphi$$

has been neglected, as well as the term $(q/T)\Delta\varphi\varphi_t$ – as done in Suppl. S5.2 – and the following notation has been introduced:

$$r_L = \left(\frac{T}{M\omega_B^2}\right)^{1/2} = \frac{c_s}{\omega_B} .$$

This is the "ion Larmor radius at electron temperature", the plasma analog of the Rossby radius in geophysical hydrodynamics.

Expressed in dimensionless units, (S5.39) has the form

$$(\Delta\varphi - \varphi)_t + V^*\varphi_x - V^*\frac{q}{T}\frac{d\ln T}{d\ln n}\varphi\varphi_x + V^*\frac{q}{T}\Delta\varphi\varphi_x + J(\varphi, \Delta\varphi) = 0 \tag{S5.40}$$

where x and y are in the units of r_L, time t is in the units of $\omega_B t$, drift velocity V^* is in the units of $\sqrt{T/M} = c_s$. The dimensional drift velocity

$$V^* = \frac{T}{M\omega_B n_0}\frac{\partial n_0}{\partial y}$$

is directed against the x axis (with the notations adopted in Fig. S5.1), since plasma density falls off towards the periphery. The direction of the drift wave corresponds to "cyclonic" rotation about the plasma column axis. This follows from the Ertel theorem (S5.33) and the definition of the ω_B vector: in the case of plasma, the cyclonic curl is directed along ω_B, that is, parallel to the magnetic field.

The propagation direction of a drift wave caused by plasma density falling off towards the periphery is the same as the direction of a Rossby wave caused by shallow water thickness falling off towards the periphery (as has been shown in

Suppl. S5.2, a Rossby wave in such a case also moves in the cyclonic direction, that is, along the system rotation).

The absolute value is

$$V^* \approx r_L^2 \cdot \frac{\omega_B}{R_n}$$

– analogous to the Rossby velocity

$$V_R \approx r_R^2 \frac{f_0}{R}$$

where

$$R_n = \frac{n_0}{\partial n_0 / \partial y}$$

is the characteristic size of the plasma density inhomogeneity. Similarly,

$$R_T = \frac{T}{\partial T / \partial y}$$

is the characteristic size of the temperature inhomogeneity, and

$$\frac{d \ln T}{d \ln n} \approx \frac{R_n}{R_T} .$$

One can easily show by linearizing (S5.40) that the drift waves propagate in the cyclonic direction (Fig. S5.1) and their phase velocity is

$$\frac{\omega}{k_x} = -\frac{V^*}{1 + k^2 r_L^2} ,$$

its limiting value being V^*.

The mechanism of drift wave generation (as well as the similar mechanism for Rossby waves) can be easily understood through the potential vortex conservation law (S5.33). Indeed, suppose that a plasma density perturbation occurs, such that some particles get displaced to where the plasma density is lower in one region, and some to where it is higher in another (Fig. S5.1). Then, according to (S5.33), the vorticity ω' must decrease in the first region and increase in the second one. In other words, an anticyclonic motion must arise in the first region and a cyclonic motion in the second one. The perturbation will then start "running" in the cyclonic direction, in quite the same manner as illustrated in Fig. 5.2.

Within the linear approximation we are discussing, drift wave propagation (the β-effect) is due only to the density gradient and does not depend on the temperature gradient. It is seen from (S5.40) that the temperature gradient can affect the velocity of drift wave propagation only in the case of a very strong nonlinearity. For instance, if

$$\frac{q\varphi}{T} \frac{d \ln T}{d \ln n} > 1 ,$$

the drift velocity will reverse its sign! This can be easily seen also from (S5.33).

Equation (S5.40) is a generalization of the well-known Hasegawa-Mima equation which does not account for such nonlinearities as $\varphi\varphi_x$ or $\Delta\varphi\varphi_x$. One can see from (S5.40) that the $\varphi\varphi_x$ nonlinearity in a plasma is due to the *temperature* gradient (in accordance with [S5.3, 4]), while its Rossby wave counterpart, hh_x, is due to the Coriolis parameter gradient. As regards the nonlinearity associated with plasma density gradient, it corresponds to the term $\Delta\varphi\varphi_x$ which is the counterpart of Δhh_x related to the gradient of shallow water thickness (see (S5.18)).

Let us now see what are the feasibility conditions for a plasma drift soliton. The condition of soliton existence (the absence of linear wave radiation) implies that, if a soliton is propagating in the same direction as linear waves, it should move faster than any such wave, that is, its velocity must satisfy

$$V = V^*\left(1 - \frac{d\ln T}{d\ln n}\cdot\frac{q\varphi}{T}\right) > V^* \ . \tag{S5.41}$$

Usually, $\frac{d\ln T}{d\ln n} > 0$. Then plasma drift solitons can only arise for $\varphi < 0$. With this (negative) sign of φ, plasma in the crossed fields B and $-\nabla\varphi$ will rotate *counter* to ion Larmor rotation so that the vortex in question will be a *cyclone*.

If, on the other hand, $d\ln T/d\ln n < 0$ (such a case seems too artificial), then a soliton is possible only for $\varphi > 0$ (an anticyclone). It is exactly this model case that is illustrated in Fig. 12.1.

There is one more possibility of soliton existence implied by (S5.41): if the amplitude is high enough so that

$$\frac{q\varphi}{T}\frac{d\ln T}{d\ln n} > 1 \ , \tag{S5.42}$$

and still $d\ln T/d\ln n > 0$, the soliton is an anticyclone and drifts in the anticyclonic direction. Considering the working conditions in magnetic traps for holding high-temperature plasma (for instance, of the Tokamak type), the requirement (S5.42) appears quite strong. In principle, it may prove much easier to satisfy this requirement if $R_T \ll R_n$ since the unfeasibly high potential $q\varphi/T > 1$ will be no longer needed.

On cyclones and anticyclones a in plasma, see [S5.5] (equations (1), (3) there and the text following (3)), also [S5.6, 7].

The material presented in this Supplement should give more insight into the analogy between Rossby waves and plasma drift waves than the discussion in Chap. 5 and help understand the limitations of this analogy.

Finally, a substantial distinction between a drift vortex and a linear drift wave should be pointed out. In the linear wave regime, plasma particles *do not drift* about the plasma column axis. In the "genuine" drift vortex regime, the particles of a plasma captured by the vortex drift along with it. Suppose a solitary vortex is propagating in the cyclonic direction. Then its velocity is higher than the drift velocity $\propto r_L^2\omega_B/R_n$. The condition of plasma capture by a drift vortex is that the velocity of plasma rotation under the influence of the crossed fields $(-\nabla\varphi,\ \boldsymbol{B})$ be higher than the drift velocity:

$$\frac{c\varphi}{aB} > \frac{r_L^2 \omega_B}{R_n}$$

or   (S5.43)

$$\frac{q\varphi}{T} > \frac{a}{R_n} \ .$$

The particle capture condition (S5.43) is necessary for soliton existence, as it was shown in Chap. 5.

It is interesting to compare the condition of particle capture (that is, of soliton existence) given by (S5.43) with the condition, discovered experimentally, under which the plasma confinement time in Tokamak-type traps becomes inversely proportional to the level of plasma drift turbulence [S5.6 – 8]. The latter condition is as follows:

$$\frac{a^2}{R_n^2} \lesssim \frac{\langle \delta n \rangle^2}{\langle n \rangle^2} \ .$$

Since in drift waves

$$\frac{\delta n}{n} \approx \frac{q\varphi}{T} \ ,$$

the above condition implies

$$\frac{q\varphi}{T} \gtrsim \frac{a}{R_n} \ ,$$

which coincides with (S5.43). Thus the validity of the drift soliton existence condition affects substantially the plasma confinement time in large thermonuclear traps (see also [S5.9, 10]).

This fact is reflected in the existence of the "drift scaling" which sums up and predicts the laws of plasma confinement in Tokamak-type thermonuclear traps [S5.8].

S8.1 On Numerical Calculations Based on the Gravitational Concept of the Origin of Galactic Spiral Structures

Some authors (see, for instance, [S8.1, 2]) report their results of numerical calculation demonstrating the formation of spiral structures in the self-gravitating stellar component of galactic disks. However, it should be noted that, as shown by Toomre [S8.3] (see also [S8.1, 2]), such spiral waves have a significant group velocity along the galaxy radius and therefore are to be damped quickly. An efficient generation mechanism is necessary to support them. But there is no *uni-versal* generation mechanism in the gravitational concept. For example, a *special* mechanism has been proposed which relies on the fact that some galaxies contain so-called "bars" (these are the SB-type galaxies). Clearly, a powerful bar is capable of generating a *dual-arm* spiral, but this is a special mechanism which cannot explain the generation of *multi-arm* spirals. The adepts of the gravitational concept are already taking into account the role of gas, not only of stars. But it is also necessary to consider velocity jumps on the rotation curves of real galaxies. This could lead to a convergence of both concepts, gravitational and hydrodynamic.

It is also of interest to mention that there is a jump of about 30 % in the rotation velocity between the central core of our Galaxy and its periphery [S8.4].

S10.1 New Results on Planetary Vortex Dynamics Obtained on the Basis of the "Shallow Water" Concept

1. It is interesting to note that the "shallow water" model proves more attractive even to those authors who only recently were inclined to be adepts of the "deep water" approach. We mean the numerical calculation reported in [S10.1, 2], based on the *shallow water* model. It demonstrates the generation of long-lived anticy-clonic vortices by unstable zonal flows with a meridional (anticyclonic) profile corresponding to that observed in the vicinity of the Jovian Great Red Spot. Several vortex generation regimes are noted. In some of them, relatively small vortices appear at first, subsequently merging to give rise to a large anticyclone, solitary over the entire range of the model which is half the planet perimeter. This vortex is assumed to be the analog of the JGRS. In other regimes, a vortex *chain* is generated, its vortices located at invariable mutual distances and showing no tendency to merge. The authors of [S10.1, 2] conclude that these chains are the analogs of the White Oval chains located in anticyclonic zones to the south of the Great Red Spot.

In both cases, the stationary state of the flow system (dissipating due to momentum loss for vortex generation) is sustained by a specially selected pumping in the form of a "Rayleigh force" $\propto \varrho v / \tau$ acting between the thick layer of Jovian atmosphere and the thin layer in which the Great Red Spot is observed (here ϱ is the medium density, v is the velocity in the upper layer, $\tau \approx 400$ days) [S10.3].

It is very important that a quite sharp cyclone-anticyclone asymmetry is also recorded in the investigations cited above for the JGRS-type vortices: while anticyclones live for more than 100 years, cyclones (with a flow profile corresponding to that observed in the cyclonic area between the JGRS and the White Ovals) are so short-lived that they are even not discussed in detail.

Yet another and quite significant effect discussed in [10.1, 2] is the topography of the "bottom" between the upper cloud layer of the Jovian atmosphere and the thick, nearly adiabatic layer which lies beneath. Such a topography can, in principle, reverse the sign of the β-effect. Indeed, since $\beta \propto (\partial/\partial y)(f/H_0)$, if there is a meridional gradient of H_0 of the same sign as the gradient of f, the sign of β can be opposite to what it was with no gradient of H_0. One may think that this effect is responsible for the fact that the Jovian White Ovals are drifting eastwards, not westwards.

2. As regards the other approach to Rossby vortices on the giant planets, based on the theory of thermoconvection in rotating "deep water", even F. Busse, its author, believes the Jovian Great Red Spot and White Ovals to be "shallow water structures" [S10.4, 5]. His argument is that, according to infrared "probing" of Jovian atmosphere, the JGRS is observed only in the very upper cloud layers where the gas pressure is significantly lower than 800 mbar [S10.6].

3. A very important result has been obtained in [S10.7] where the following crucial fact is discussed. The initial equations for shallow water, even in the absence of the β-effect, contain asymmetry between cyclones and anticyclones (similar to that which "permits" elevation solitons and "forbids" depression solitons in the case of no rotation). If, however, the quasi-geostrophic approximation is used (valid with no free surface or with $a < r_R$) – that is, if one uses the Charney-Obukhov equation (5.1) – then the asymmetry vanishes. This proves that the quasi-geostrophic approach does not reflect the actual situation in full. In this connection, it is shown in [S10.7] that the above asymmetry is recovered, even in the absence of the β-effect, if the nonlinearity is sufficiently strong, that is, if the value of h is not small with respect to unity. This new result is due to the fact that the generalized geostrophic equation (S5.13) takes into account an additional term, cubic with respect to h, namely

$$\frac{g^* H_0}{f_0} h \cdot J(h, \Delta h) \ .$$

In the derivation of (S5.13) we have neglected this term (see the conditions (S5.14)), assuming for simplicity $h \ll 1$. In reality, however, the value of h cannot be too small: since the vortex must carry trapped particles, then, as shown earlier (see, for instance, Chap. 5),

$$h \gg \frac{a}{R} \ .$$

This is why the new term, which does not contain β, is, in principle, essential. This is very important because a cubic term with respect to h violates the cyclone-anticyclone symmetry. Several effects which demonstrate the stability of anticyclones and the decay of cyclones are described in [S10.7].

4. Sometimes (see [S10.8]) attention is drawn to the fact that cyclone-anticyclone asymmetry is inherent even in the smaller vortices of Jovian atmosphere: anticyclones are markedly prevailing even among the vortices with $a < r_R$. We are inclined to explain this as follows. In the case of an anticyclone, the force related to the gradient of hydrostatic pressure equilibrates the *difference* between

the Coriolis acceleration and the relative acceleration due to the proper rotation of the vortex, while in the case of a cyclone the gradient force equilibrates the *sum* of the two accelerations. Cyclonic profiles are therefore steeper and, consequently, less stable. As a result, even small cyclones decay sooner than small anticyclones.

However, it should be recalled that, speaking of cyclone-anticyclone asymmetry, we mean only *long-lived* vortices while the small vortices ($a < r_R, r_i$) are short-lived. For example, the small vortices which constitute the cyclone-anticyclone "street" at about 41°S in the Jovian atmosphere, live on the average for two or three months (see [S10.9] and Fig. 6 there, also [S10.10] and Figs. 1, 8 there). Therefore such vortices are not included in our classification at all.

References

Chapter 2

2.1   C.G. Rossby: J. Marine Res. **2**, 38 (1939)
2.2   C.G. Rossby: Quart. J. Meteor. Soc. Suppl. **66**, 68 (1940)
2.3   C.G. Rossby: J. Marine Res. **7**, 175 (1948)
2.4   B. Haurwitz: J. Marine. Res. **3**, 254 (1940)
2.5   E.N. Lorenz: The Nature and Theory of the General Circulation of the Atmosphere (WMO, Geneva 1967)
2.6   H. Stommel: The Gulf Stream, 2nd ed. (University of California Press, Berkeley 1965)
2.7   H. Stommel, K. Yoshida: Kuroshio. Its Physical Aspects (University of Tokyo Press, 1972)
2.8   H.D.I. Abarbanel, W.R. Yong (eds.): General Circulation of the Ocean (Springer, Berlin, Heidelberg 1987)
2.9   J. Pedlosky: Geophysical Fluid Dynamics (Springer, New York 1987)
2.10 P.H. Le Blond, L.A. Mysak: Waves in the Ocean (Elsevier, New York 1978)
2.11 H.J. Lugt: Vortex Flow in Nature and Technology (Wiley, New York 1983)
2.12 A.K. Gupta, D.G. Lilley, N. Syred: Swirl Flows (Abacus, England 1987)
2.13 L. Bengtsson, J. Lighthill (eds.): Intense Atmospheric Vortices (Springer, New York 1982)
2.14 B. Hoskins, R. Pearce (eds.): Large Scale Dynamical Processes in the Atmosphere (Academic Press, New York 1983)
2.15 A.P. Khain, G.G. Sutyrin: Tropicheskiye tsiklony i ikh vzaimodeistviye s okeanom (Tropical Cyclones and Their Interaction with the Ocean) (Gidrometeoizdat, Leningrad 1983)
2.16 R.A. Anthes: Tropical Cyclones, Their Evolution, Structure, and Effects (Amer. Meteorol. Soc. 1982)
2.17 M. van Dyke: An Album of Fluid Motion (Parabolic Press, Stanford 1982)
2.18 A.E. Gill: Atmosphere-Ocean Dynamics, Vols. 1,2 (Academic Press, New York 1982)
2.19 P.B. Rhines: J. Fluid Mech. **69**, 417 (1975)
2.20 P.B. Rhines: Ann. Rev. Fluid Mech. **11**, 401 (1979)
2.21 G.P. Williams: Nature, **257**, 778 (1975)
2.22 G.P. Williams: J. Atmos. Sci. **35**, 1399 (1978)
2.23 G.P. Williams: J. Atmos. Sci. **36**, 932 (1979)
2.24 R.H. Kraichan: Repts. Progr. Phys. **43**, 547 (1979)
2.25 V.D. Larichev: Dokl. AN SSSR **298**, 971 (1988)
2.26 V.P. Starr: Physics of Negative Viscosity Phenomena (McGraw-Hill, New York 1968)
2.27 G.B. Whitham: Linear and Nonlinear Waves (Wiley, New York 1974)
2.28 T. Gehrels (ed.): Jupiter (University of Arizona Press, Tucson 1976)
2.29 G.E. Hunt, P. Moore: Ann. Rev. Earth Planet Sci. **11**, 415 (1983)
2.30 T. Gehrels, M.S. Matthews (eds.): Saturn (University of Arizona Press, Tucson 1984)
2.31 G.E. Hunt, P. Moore: Jupiter (Rand McNally, Chicago 1981)
2.31' R. Hooke: Phil. Trans. Roy. Soc. **1**, A2 (1666)
2.32 G.E. Hunt, P. Moore: Saturn (Roy. Astron. Soc. and Rand McNally, London 1982)
2.33 B.A. Smith, R.A. Soderblom, T.V. Johnson, et al.: Science **204**, 951 (1979)
2.34 B.A. Smith, R.A. Soderblom, R. Batson, et al.: Science **215**, 504 (1982)

2.35 R.F. Beebe, T.A. Hockey: Icarus **65**, 86 (1986)
2.36 A. Hatzes, D.D. Wenkert, A.P. Ingersoll, G.E. Danielson: J. Geophys. Res. **A86**, 8745 (1981)
2.37 A.P. Ingersoll, D. Pollard: Icarus **52**, 62 (1982)
2.38 A.P. Ingersoll, R.F. Beebe, S.A. Collins: Nature **280**, 773 (1979)
2.39 A.P. Ingersoll: Sci. Amer. **245(6)**, 66 (1981)
2.40 A.P. Ingersoll, R.L. Miller: Icarus **65**, 370 (1986)
2.41 M.M. McLow, A.P. Ingersoll: Icarus **65**, 353 (1986)
2.42 J.L. Mitchell, R.F. Beebe, A.P. Ingersoll, G.W. Garneau: J. Geophys. Res. **A86**, 8751 (1981)
2.43 C.W. Hord, R.A. West, K.E. Simmons: Science **206**, 956 (1979)
2.44 S.S. Limaye: Icarus **65**, 335 (1986)
2.45 V.R. Eshleman, G.L. Tyler, G.E. Wood, G.F. Lindal, J.D. Anderson, G.S. Levy: Science **204**, 976 (1979)
2.46 L.A. Sromovsky, H.E. Revercomb, R.J. Krauss, V.E. Suomi: J. Geophys. Res. **A88**, 8650 (1983)
2.47 D.A. Godfrey, V. Moore: Icarus **68**, 313 (1986)
2.48 A.P. Ingersoll, R.F. Beebe, B.J. Conrath, G.E. Hunt: in [2.30], p. 195
2.49 D.J. Tritton, P.A. Davis: in Hydrodynamic Instabilities and Transition To Turbulence, ed. by H. Swinney, J. Gollub (Springer, New York 1981), Chap. 8
2.50 A.M. Obukhov: Turbulentnost' i dinamika atmosfery (Turbulence and Atmospheric Dynamics) (Gidrometeoizdat, Leningrad 1988)
2.51 F.V. Dolzhansky: Izv. Atm. Ocean. Phys. **17**, 413 (1981)
2.52 A.M. Obukhov, M.V. Kurgansky, M.S. Tatarskaya: Sov. Meteor. Hydrol. **10**, 1 (1984)
2.53 R.A. Madden: Rev. Geophys. and Space Phys. **17**, 1935 (1979)
2.54 D. Andrews: Nature **310**, 185 (1984)
2.55 V.M. Kamenkovich, M.N. Koshlyakov, A.S. Monin: Sinopticheskiye vikhri v okeane (Synoptic Vortices in the Ocean) (Gidrometeoizdat, Leningrad 1987)
2.56 A.S. Monin, M.N. Koshlyakov: in Nelineinye Volny (Nonlinear Waves), ed. by A.V. Gaponov-Grekhov (Nauka, Moscow 1979), p. 258
2.57 A.R. Robinson (ed): Eddies in Marine Sciences (Springer, New York 1983)
2.58 J.C.J. Nihoul, B.M. Jamart (eds.): Mesoscale/Synoptic Coherent Structures in Geophysical Turbulence (Elsevier, Amsterdam 1989)
2.59 D.C. Smith IV, R.O. Reid: J. Phys. Oceanogr. **12**, 244 (1982)
2.60 K.N. Fedorov (ed.): Vnutritermoklinnye vikhri v okeane (Intrathermoclinic Vortices in the Ocean) (Acad. Sci. USSR, P.P. Shirshov Institute of Oceanology, Moscow 1986)
2.61 S.E. McDowell, H.T. Rossby: Science **202**, 1085 (1978)
2.62 L. Armi, D. Hebert, N. Oakey: Nature **333**, 649 (1988)
2.63 J. Marshall: Nature **333**, 594 (1988)
2.64 A.G. Kostyanoi, G.I. Shapiro: in [2.60], p. 120
2.65 G.I. Shapiro: in [2.60], pp. 56, 71, 79
2.66 J.C. McWilliams: Rev. Geophys. **23**, 165 (1985)
2.67 J.C. McWilliams, P.R. Gent, N.J. Norton: J. Phys. Oceanogr. **16**, 838 (1986)
2.68 A.G. Kostyanoi, I.M. Belkin: in [2.58], p. 821
2.69 G.P. Williams: Adv. Geophys. **28A**, 381 (1985)
2.70 A.M. Fridman, V.L. Polyachenko: Physics of Gravitating Systems, Vols. 1,2 (Springer, New York 1984)
2.71 J. Jeans: Astronomy and Cosmogony (Cambridge University Press, Cambridge 1928)
2.72 W. Heisenberg, C.F. Weizsäcker: Z. Phys. **125**, 290 (1948)
2.73 S. Chandrasekhar, E. Fermi: Astrophys. J. **118**, 113 (1953)
2.74 B. Lindblad: Stock. Obs. Ann. **13**, 10 (1941)
2.75 C. Lin, F. Shu: Astrophys. J. **140**, 646 (1964)
2.76 K. Rohlfs: Lectures on Density Wave Theory (Springer, Berlin, Heidelberg, New York 1977)
2.77 W.C. Saslaw: Gravitational Physics of Stellar and Galactic Systems (Cambridge University Press, Cambridge 1987)

2.78   L.S. Marochnik, A.A. Suchkov: Galaktika (Galaxy) (Nauka, Moscow 1984)
2.79   E. Athanassoula: Phys. Repts. **114**, 314 (1985)
2.80   A. Toomre: in Structure and Evolution of Normal Galaxies, ed. by S.M. Fall, D. Linden-Bell (Cambridge University Press, Cambridge 1981), p. 111
2.81   H. Arp: IEEE Trans. Plasma Sci. **PS-14**, 748 (1986)
2.82   A.M. Fridman: Sov. Phys. Usp. **21**, 536 (1978)
2.83   A.G. Morozov: Sov. Astron. Lett. **3**, 103 (1977)
2.84   A.G. Morozov: Sov. Astron. **23**, 278 (1979)
2.85   D.B. Sanders, P.M. Solomon, N.Z. Scoville: Astrophys. J. **276**, 182 (1984)
2.86   E.D. Pavlovskaya, A.A. Suchkov: Astronom. Zh. **57**, 280 (1980)
2.87   U.A. Haud: Pis'ma v Astronom. Zh. **5**, 124 (1979)
2.88   V.C. Rubin, W.K. Ford: Astrophys. J. **159**, 379 (1970)
2.89   R.R. Sinha: Astron. Astrophys. **69**, 227 (1978)
2.90   J.W. Goad: Astrophys. J. Suppl. Ser. **32**, 89 (1976)
2.91   V.C. Rubin, D. Burstein, W.K. Ford Jr., N. Tronnard: Astrophys. J. **81**, 289 (1985)
2.92   A.M. Fridman: in Dynamics of Astrophysical Disks, ed. by J.A. Sellwood (Cambridge University Press, Cambridge 1989), p. 185

Chapter 3

3.1   A.M. Fridman: Vestnik AN SSSR **6**, 18 (1987)
3.2   A.M. Fridman: in Dynamics of Astrophysical Disks, ed. by J.A. Sellwood (Cambridge University Press, Cambridge 1989), p. 185
3.3   G.F. Chew, M.L. Goldberger, F.E. Low: Proc. Roy. Soc. **236A**, 112 (1956)
3.4   J. Pedlosky: Geophysical Fluid Dynamics (Springer, New York 1987)
3.5   P.H. Le Blond, L.A. Mysak: Waves in the Ocean (Elsevier, New York 1978)
3.6   A.P. Ingersoll, D. Pollard: Icarus **52**, 62 (1982)
3.7   S.S. Limaye: Icarus **65**, 335 (1986)
3.8   L.A. Sromovsky, H.E. Revercomb, R.J. Krauss, V.E. Suomi: J. Geophys. Res. **A88**, 8650 (1983)
3.9   D.A. Godfrey, V. Moore: Icarus **68**, 313 (1986)
3.10   A.P. Ingersoll, R.F. Beebe, B.J. Conrath, G.E. Hunt: in Saturn, ed. by T. Gehrels, M.S. Matthews (University of Arizona Press, Tucson 1984), p. 195
3.11   A.M. Fridman, V.L. Polyachenko: Physics of Gravitating Systems, Vols. 1,2 (Springer, New York 1984)

Chapter 4

4.1   L.D. Landau, E.M. Lifshitz: Gidrodinamika (Hydrodynamics), 3rd ed. (Nauka, Moscow 1986)
4.2   S. Chandrasekhar: Hydrodynamics and Hydrodynamic Stability (Clarendon, Oxford 1961)
4.3   A.V. Timofeyev: in Voprosy teorii plasmy (Topics in Plasma Theory) No. 17, ed. by B.B. Kadomtsev (Energoatomizdat, Moscow 1989), p. 157
4.4   L.A. Ostrovsky, Yu.A. Stepanyants: Rev. Geophys. **27**, 293 (1989)
4.5   L.D. Landau: Dokl. AN SSSR **44**, 151 (1944)
4.6   S.V. Bazdenkov, O.P. Pogutse: JETP Lett. **37**, 375 (1983)
4.7   S.V. Antipov, M.V. Nezlin, V.K. Rodionov, E.N. Snezhkin, A.S. Trubnikov: JETP Lett. **37**, 378 (1983)
4.8   M.V. Nezlin: Sov. Phys. Usp. **29**, 807 (1986)

Chapter 5

5.1   L.D. Landau, E.M. Lifshitz: Gidrodinamika (Hydrodynamics), 3rd revised ed. (Nauka, Moscow 1986)
5.2   J.G. Charney: Geophys. Public. **17**, 3 (1948)
5.3   A.M. Obukhov: Izv. AN SSSR, Geografia i Geofizika, **13**, 281 (1949)
5.4   J. Pedlosky: Geophysical Fluid Dynamics (Springer, New York 1987)
5.5   G.K. Batchelor: Introduction to Fluid Dynamics (Cambridge University Press, New York 1967)
5.6   V.D. Larichev, G.M. Reznik: Dokl. AN SSSR **231**, 1077 (1976)
5.7   A. Hasegawa, K. Mima: Phys. Fluids **21**, 87 (1978)
5.8   A. Hasegawa, S.G. McLennan, Y. Kodama: Phys. Fluids **22**, 2122 (1979)
5.9   A. Hasegawa: Adv. Phys. **34**, 1 (1985)
5.10 M. Makino, T. Kamimura, T. Tainuti: J. Phys. Soc. Japan **50**, 990 (1981)
5.11 J.G. Charney, G.R. Flierl: in Evolution of Physical Oceanography, ed. by B.A. Warren, C. Wunsch (MIT Press, Cambridge 1981)
5.12 G.R. Flierl: POLYMODE News **62**, 1 (1979)
5.13 D.L.T. Anderson, P.D. Killworth: Deep-Sea Res. **26A**, 1033 (1979)
5.14 T. Matsuura, T. Yamagata: J. Phys. Oceanogr. **12**, 440 (1982)
5.15 G.P. Williams, T. Yamagata: J. Atm. Sci. **41**, 453 (1984)
5.16 G.P. Williams: Adv. Geophys. **28A**, 381 (1985)
5.17 G.P. Williams, R.J. Wilson: J. Atm. Sci. **45**, 207 (1988)
5.18 E.N. Mikhailova, N.B. Shapiro: Izv. Atm. Ocean Phys **16**, 587 (1980)
5.19 V.I. Petviashvili: Sov. Phys. JETP Lett. **32**, 619 (1980)
5.20 Yu.A. Danilov, V.I. Petviashvili: in Itogi nauki i tekhniki, ser. Fizika plasmy (Advances in Science and Technology, Ser. Plasma Physics) No.4, ed. by V.D. Shafranov (VINITI, Moscow 1983), p. 5
5.21 N.N. Romanova, V.Yu. Tseitlin: Izv. Atm. Ocean Phys **20**, 85 (1984)
5.22 J. Nycander: Phys. Scripta **39**, 758 (1989)
5.23 J. Nycander: Phys. Fluids **B1**, 1788 (1989)
5.24 J. Nycander: in Proc. IV Intern. Workshop on Nonlinear and Turbulent Proc. in Physics, ed. by A.G. Sitenko, V.E. Zakharov, V.M. Chernousenko (Naukova Dumka, Kiev 1989) Vol. 1, p. 399
5.25 J. Nycander: J. Plasma Phys. **39**, 413 (1988)
5.26 A.C. Scott, F.Y.E. Chu, D.W. McLaughlin: Proc. IEEE **61**, 1443 (1973)
5.27 P.H. Le Blond, L.A. Mysak: Waves in the Ocean (Elsevier, New York 1978)
5.28 A.M. Obukhov, M.V. Kurgansky, M.S. Tatarskaya: Sov. Meteor. Hydrol. **10**, 1 (1984)
5.29 R.A. Madden: Rev. Geophys. and Space Phys. **17**, 1935 (1979)
5.30 D. Andrews: Nature **310**, 185 (1984)
5.31 L.A. Sromovsky, H.E. Revercomb, R.J. Krauss, V.E. Suomi: J. Geophys. Res. **A88**, 8650 (1983)
5.32 D.A. Godfrey, V. Moore: Icarus **68**, 313 (1986)
5.33 A.P. Ingersoll, R.F. Beebe, B.J. Conrath, G.E. Hunt: in Saturn, ed. by T. Gehrels, M.S. Matthews (University of Arizona Press, Tucson 1984), p. 195
5.34 G.F. Chew, M.L. Goldberger, F.E. Low: Proc. Roy. Soc. **236A**, 112 (1956)
5.35 N.J. Zabusky, M.D. Kruskal: Phys. Rev. Lett. **15**, 240 (1965)
5.36 G.R. Flierl: J. Phys. Oceanogr. **7**, 365 (1977)
5.37 S.V. Antipov, M.V. Nezlin, V.K. Rodionov, A.Yu. Rylov, E.N. Snezhkin, A.S. Trubnikov, A.V. Khutoretsky: Sov. J. Plasma Phys. **14**, 648 (1988)
5.38 J.W. Miles: J. Fluid Mech. **106**, 186 (1981)
5.39 R.K. Dodd, J.C. Eilbeck, J.D. Gibbon, H.C. Morris: Solitons and Nonlinear Wave Equations (Academic Press, London 1982)
5.40 H. Lamb: Hydrodynamics (Dover, New York 1945)

5.41 V.D. Larichev, G.M. Reznik: Dokl. AN SSSR **264**, 229 (1982)
5.42 P.M. Rizzoli: Adv. Geophys. **24**, 147 (1983)
5.43 M.V. Nezlin: Sov. Phys. Usp. **29**, 807 (1986)
5.44 G.B. Whitham: Linear and Nonlinear Waves (Wiley, New York 1974)
5.45 M.J. Lighthill: Waves in Fluids (Cambridge University Press, New York 1978)
5.46 B.B. Kadomtsev: Kollektivnye yavleniya v plasme (Collective Phenomena in Plasma) (Nauka, Moscow 1988)
5.47 B.B. Kadomtsev, V.I. Karpman: Soc. Phys. Usp. **14**, 40 (1971)
5.48 V.P. Pavlenko, V.V. Taranov: Fizika Plazmy **10**, 1303 (1984)
5.49 V.I. Petviashvili, A.P. Smirnov: Dokl. AN SSSR **277**, 88 (1984)
5.50 G.G. Sutyrin: Dokl. AN SSSR **280**, 1101 (1985)
5.51 G.G. Sutyrin: in Vnutritermoklinnye vikhri v okeane (Intrathermoclinic Vortices in the Ocean), ed. by K.N. Fedorov (Acad. Sci. USSR, P.P. Shirshov Institute of Oceanology, Moscow 1986), p. 93
5.52 G.G. Sutyrin: Izv. AN SSSR, Mekhanika Zhidkosti i Gaza **4**, 119 (1985)
5.53 G.G. Sutyrin: Dokl. AN SSSR **296**, 1076 (1987)
5.54 G.G. Sutyrin, I.G. Yushina: Dokl. AN SSSR **299**, 580 (1988)
5.55 Yu.A. Stepanyants, private communication

Chapter 6

6.1   S.V. Antipov, M.V. Nezlin, E.N. Snezhkin, A.S. Trubnikov: JETP Lett. **33**, 351 (1981)
6.2   S.V. Antipov, M.V. Nezlin, E.N. Snezhkin, A.S. Trubnikov: Sov. Phys. JETP **55**, 85 (1982)
6.3   S.V. Antipov, M.V. Nezlin, V.K. Rodionov, E.N. Snezhkin, A.S. Trubnikov: JETP Lett. **35**, 645 (1982)
6.4   M.V. Nezlin, E.N. Snezhkin, A.S. Trubnikov: JETP Lett. **36**, 234 (1982)
6.5   S.V. Antipov, M.V. Nezlin, V.K. Rodionov, E.N. Snezhkin, A.S. Trubnikov: Sov. Phys. JETP **57**, 786 (1983)
6.6   S.V. Antipov, M.V. Nezlin, E.N. Snezhkin, A.S. Trubnikov: in Nonlinear and Turbulent Processes in Physics, ed. by R.Z. Sagdeev (Gordon & Breach, New York 1984), Vol. 2, p. 665
6.7   S.V. Antipov, M.V. Nezlin, A.S. Trubnikov: JETP Lett. **41**, 30 (1985)
6.8   S.V. Antipov, M.V. Nezlin, E.N. Snezhkin, A.S. Trubnikov: Sov. Phys. JETP **62**, 1097 (1985)
6.9   S.V. Antipov, M.V. Nezlin, E.N. Snezhkin, A.S. Trubnikov: Nature **323**, 238 (1986)
6.10 M.V. Nezlin, E.N. Snezhkin: in Nelineinye volny (Nonlinear Waves), ed. by A.V. Gaponov-Grekhov, M.I. Rabinovich (Nauka, Moscow 1987), p. 67
6.11 S.V. Antipov, M.V. Nezlin, V.K. Rodionov, A.Yu. Rylov, E.N. Snezhkin, A.S. Trubnikov, A.V. Khutoretsky: Sov. J. Plasma Phys. **14**, 648 (1988)
6.12 M.V. Nezlin, E.N. Snezhkin, A.S. Trubnikov: in Proc. Intern. Conf. on Plasma Phys. Kiev 1987, Inv. Papers, ed. by A.G. Sitenko (World Scientific, Singapore 1987), Vol. 2, p. 1184
6.13 D. Fultz, T.S. Murty: J. Atm. Sci. **25**, 779 (1968)
6.14 V.I. Petviashvili: Sov. Phys. JETP Lett. **32**, 619 (1980)
6.15 A.V. Khutoretsky: Izv. AN SSSR, Fizika Atmosfery i Okeana **22**, 344 (1986)
6.16 M.V. Nezlin: Sov. Phys. Usp. **29**, 807 (1986)
6.17 J. Pedlosky: Geophysical Fluid Dynamics (Springer, New York 1987)
6.18 A.G. Morozov, M.V. Nezlin, E.N. Snezhkin, A.M. Fridman: Sov. Phys. Usp. **28**, 101 (1985)
6.19 M.V. Nezlin, A.Yu. Rylov, E.N. Snezhkin, A.S. Trubnikov: Sov. Phys. JETP **65**, 1 (1987)

Chapter 7

7.1   S.V. Antipov, M.V. Nezlin, E.N. Snezhkin, A.S. Trubnikov: Sov. Phys. JETP **55**, 85 (1982)
7.2   M.V. Nezlin: Sov. Phys. Usp. **29**, 807 (1986)

212   References

7.3   S.V. Antipov, M.V. Nezlin, V.K. Rodionov, A.Yu. Rylov, E.N. Snezhkin, A.S. Trubnikov,
      A.V. Khutoretsky: Sov. J. Plasma Phys. **14**, 648 (1988)
7.4   S. Chandrasekhar: Hydrodynamics and Hydrodynamic Stability (Clarendon Press, Oxford
      1961)
7.5   Ya.B. Zel'dovich, P.I. Kolykhalov: Sov. Phys. Dokl. **27**, 699 (1982)
7.6   M. Rabaud, Y. Couder: J. Fluid Mech. **136**, 291 (1983)
7.7   J.M. Chomaz, M. Rabaud, C. Basdevant, Y. Couder: J. Fluid Mech. **187**, 115 (1988)
7.8   H. Niino, N. Misawa: J. Atm. Sci. **41**, 1992 (1984)
7.9   Yu.L. Chernous'ko: Izv. Atm. Ocean Phys. **16**, 285 (1980)
7.10 F.V. Dolzhansky: Izv. Atm. Ocean. Phys. **17**, 413 (1981)
7.11 J. Sommeria, S.D. Meyers, H.L. Swinney: Nature **331**, 689 (1988)
7.12 S.D. Meyers, J. Sommeria, H.L. Swinney: Physica **D37**, 515 (1989)
7.13 S.V. Antipov, M.V. Nezlin, A.S. Trubnikov: JETP Lett. **41**, 30 (1985)
7.14 S.V. Antipov, M.V. Nezlin, E.N. Snezhkin, A.S. Trubnikov: Sov. Phys. JETP **62**, 1097 (1985)
7.15 S.V. Antipov, M.V. Nezlin, E.N. Snezhkin, A.S. Trubnikov: Nature **323**, 238 (1986)
7.16 M.V. Nezlin, E.N. Snezhkin: in Nelineinye volny (Nonlinear Waves), ed. by A.V. Gaponov-
      Grekhov, M.I. Rabinovich (Nauka, Moscow 1987), p. 67
7.17 M.V. Nezlin, E.N. Snezhkin, A.S. Trubnikov: in Proc. Intern. Conf. on Plasma Phys. Kiev
      1987, Inv. Papers, ed. by A.G. Sitenko (World Scientific, Singapore 1987), Vol. 2, p. 1184
7.18 J.L. Mitchell, R.F. Beebe, A.P. Ingersoll, G.W. Garneau: J. Geophys. Res. **A86**, 8751 (1981)
7.19 G.P. Williams: Nature, **257**, 778 (1975)
7.20 G.P. Williams: J. Atmos. Sci. **35**, 1399 (1978)
7.21 G.P. Williams: J. Atmos. Sci. **36**, 932 (1979)
7.22 G.P. Williams, R.J. Wilson: J. Atmos. Sci. **45**, 207 (1988)
7.23 B.A. Smith, R.A. Soderblom, T.V. Johnson, et al.: Science **204**, 951 (1979)
7.24 B.A. Smith, R.A. Soderblom, R. Batson, et al.: Science **215**, 504 (1982)
7.25 C.W. Hord, R.A. West, K.E. Simmons: Science **206**, 956 (1979)
7.26 S.S. Limaye: Icarus **65**, 335 (1986)
7.27 V.R. Eshleman, G.L. Tyler, G.E. Wood, G.F. Lindal, J.D. Anderson, G.S. Levy: Science **204**,
      976 (1979)
7.28 G.E. Hunt, P. Moore: Jupiter (Rand McNally, Chicago 1981)
7.29 P.J. Conrath, P.J. Gierarsch, N. Nath: Icarus **48**, 256 (1981)
7.30 J. Pedlosky: Geophysical Fluid Dynamics (Springer, New York 1987)
7.31 G.S. Golitsyn: Icarus **13**, 1 (1970)
7.32 S.A. Maslou: in Hydrodynamic Instabilities and Transition To Turbulence, ed. by H. Swinney,
      J. Gollub (Springer, New York 1981), Chap. 7
7.33 Ho Chih-Ming, P. Huerre: Ann. Rev. Fluid Mech. **16**, 365 (1984)
7.34 M.V. Nezlin: JETP Lett. **34**, 77 (1981)
7.35 M.V. Nezlin: Sov. Astron. Lett. **10**, 221 (1984)
7.36 E.J. Reese, B.A. Smith: Icarus **9**, 474 (1968)
7.37 A.P. Ingersoll, R.L. Miller: Icarus **65**, 370 (1986)
7.38 M.M. McLow, A.P. Ingersoll: Icarus **65**, 353 (1986)
7.39 G.P. Williams: Adv. Geophys. **28A**, 381 (1985)
7.40 G.P. Williams, T. Yamagata: J. Phys. Oceanogr. **12**, 440 (1982)
7.41 G.G. Sutyrin: in Vnutritermoklinnye vikhri v okeane (Intrathermoclinic Vortices in the Ocean),
      ed. by K.N. Fedorov (Acad. Sci. USSR, P.P. Shirshov Institute of Oceanology, Moscow 1986),
      p. 93
7.42 P.L. Read, R. Hide: Nature **302**, 126 (1983)
7.43 P.L. Read, R. Hide: Nature **308**, 45 (1984)
7.44 T.E. Dowling, A.P. Ingersoll: J. Atmos. Sci. **45**, 1380 (1988)
7.45 S.V. Antipov, M.V. Nezlin, E.N. Snezhkin, A.S. Trubnikov: Nature **343**, 517 (1990)
7.46 J. Sommeria, S.D. Meyers, H.L. Swinney: Nature **337**, 58 (1989)
7.47 A.M. Obukhov, M.V. Kurgansky, M.S. Tatarskaya: Sov. Meteor. Hydrol. **10**, 1 (1984)

7.48 D. Andrews: Nature **310**, 185 (1984)
7.49 V.D. Larichev, G.M. Reznik: Dokl. AN SSSR **264**, 229 (1982)
7.50 M.V. Nezlin, E.N. Snezhkin, A.S. Trubnikov: JETP Lett. **36**, 234 (1982)
7.51 P. Ripa: J. Fluid Mech. **126**, 463 (1983)
7.52 A.P. Ingersoll, D. Pollard: Icarus **52**, 62 (1982)
7.53 T. Matsuura, T. Yamagata: J. Phys. Oceanogr. **12**, 440 (1982)
7.54 A.V. Timofeyev: in Voprosy teorii plasmy (Topics in Plasma Theory) No. 17, ed. by B.B. Kadomtsev (Energoatomizdat, Moscow 1989), p. 157
7.55 P.G. Drazin, W.H. Reid: Hydrodynamic Stability (Cambridge University Press, Cambridge 1981)
7.56 Yu.A. Stepanyants, A.L. Fabrikant: Usp. Fiz. Nauk **159**, No. 1, 83 (1989)
7.57 J. Sommeria, S.D. Meyers, H.L. Swinney: in Nonlinear Topics in Ocean Physics, ed. by A. Osborn (North Holland, Amsterdam 1989)
7.58 M.E. McIntyre: Ann. Geophys., Suppl. to Vol. **9**, 338 (1991)
7.59 A. Plumb: Nature **347**, 20 (1990)
7.60 R. Pool: Nature **350**, 451 (1990)
7.61 S.V. Antipov, M.V. Nezlin, V.K. Rodionov, E.N. Snezhkin, A.S. Trubnikov: JETP Lett. **37**, 378 (1983)
7.62 S.M. Churilov, I.G. Shukhman: Preprint No. 15-90 (SibIZMIR, 1990)

Chapter 8

8.1   A.G. Morozov, M.V. Nezlin, E.N. Snezhkin, A.M. Fridman: Sov. Phys. Usp. **28**, 101 (1985)
8.2   M.V. Nezlin, A.Yu. Rylov, E.N. Snezhkin, A.S. Trubnikov: Sov. Phys. JETP **65**, 1 (1987)
8.3   A.G. Morozov, M.V. Nezlin, E.N. Snezhkin, A.M. Fridman: JETP Lett. **39**, 613 (1984)
8.4   A.M. Fridman, A.G. Morozov, M.V. Nezlin, E.N. Snezhkin: Phys. Lett. **A109**, 228 (1985)
8.5   A.M. Fridman, A.G. Morozov, M.V. Nezlin, I.I. Pasha, V.L. Polyachenko, A.Yu. Rylov, E.N. Snezhkin, A.S. Trubnikov: in Observational Evidence of Activity in Galaxies, ed. by E.Ye. Khachikyan (Intern. Astron. Union, Burakan 1987), p. 147
8.6   M.V. Nezlin, V.L. Polyachenko, E.N. Snezhkin, A.S. Trubnikov, A.M. Fridman: Sov. Astron. Lett. **12**, 213 (1986)
8.7   M.V. Nezlin: Vestnik AN SSSR **6**, 28 (1987)
8.8   L. Bengtsson, J. Lighthill (eds.): Intense Atmospheric Vortices (Springer, New York 1982)
8.9   B. Hoskins, R. Pearce (eds.): Large Scale Dynamical Processes in the Atmosphere (Academic Press, New York 1983)
8.10 A.P. Khain, G.G. Sutyrin: Tropicheskiye tsiklony i ikh vzaimodeistviye s okeanom (Tropical Cyclones and Their Interaction with the Ocean) (Gidrometeoizdat, Leningrad 1983)
8.11 R.A. Anthes: Tropical Cyclones, Their Evolution, Structure, and Effects (Amer. Meteorol. Soc. 1982)
8.12 A.M. Fridman: Sov. Phys. Usp. **21**, 536 (1979)
8.13 A.G. Morozov: Sov. Astron. Lett. **3**, 103 (1977)
8.14 A.G. Morozov: Sov. Astron. **23**, 278 (1979)
8.15 A.M. Fridman: in Dynamics of Astrophysical Disks, ed. by J.A. Sellwood (Cambridge University Press, Cambridge 1989), p. 185
8.16 A.G. Morozov, M.V. Nezlin, E.N. Snezhkin, Yu.M. Torgashin, A.M. Fridman: Astronomicheskiy Tsirkulyar **1414**, 1, 4, 7 (1986)
8.17 V.L. Afanasyev, P.V. Bayev, A.M. Fridman, I.N. Ivchenko, Yu.N. Makov, A.G. Morozov, M.V. Nezlin, I.I. Pasha, V.L. Polyachenko, A.Yu. Rylov, E.N. Snezhkin, Yu.M. Torgashin, A.S. Trubnikov, A.V. Zasov: in Evolution of Galaxies, ed. by J. Palous (Astron. Inst. of the Czechosl. Acad. Sci. Prague 1987), p. 301
8.18 S.V. Bazdenkov, N.N. Morozov, O.P. Pogutse: Preprint 4298/6 (Kurchatov Institute for Atomic Energy, Moscow 1986)

8.19 S.V. Bazdenkov, N.N. Morozov, O.P. Pogutse: Preprint 4250/6 (Kurchatov Institute for Atomic Energy, Moscow 1986)
8.20 A.J. Faller: J. Fluid Mech. **15**, 560 (1963)
8.21 R. Kobayashi, Y. Kohama, Ch. Takamadata: Acta Mech. **35**, 71 (1980)
8.22 J. Sommeria, S.D. Meyers, H.L. Swinney: Nature **331**, 689 (1988)
8.23 S.D. Meyers, J. Sommeria, H.L. Swinney: Physica **D37**, 515 (1989)
8.24 M. Rabaud, Y. Couder: J. Fluid Mech. **136**, 291 (1983)
8.25 J.M. Chomaz, M. Rabaud, C. Basdevant, Y. Couder: J. Fluid Mech. **187**, 115 (1988)
8.26 S.V. Antipov, M.V. Nezlin, E.N. Snezhkin, A.S. Trubnikov: Sov. Phys. JETP **62**, 1097 (1985)
8.27 S.V. Antipov, M.V. Nezlin, V.K. Rodionov, A.Yu. Rylov, E.N. Snezhkin, A.S. Trubnikov, A.V. Khutoretsky: Sov. J. Plasma Phys. **14**, 648 (1988)
8.28 M.V. Nezlin, E.N. Snezhkin, A.S. Trubnikov: in Proc. Intern. Conf. on Plasma Phys. Kiev 1987, Inv. Papers, ed. by A.G. Sitenko (World Scientific, Singapore 1987), Vol. 2, p. 1184
8.29 F.V. Dolzhansky: Izv. AN SSSR, Fizika Atmosfery i Okeana **23**, 348 (1987)
8.30 E.B. Gledzer, F.V. Dolzhansky, A.M. Obukhov: Sistemy gidrodinamicheskogo tipa i ikh primenenie (Systems of Hydrodynamic Type and Their Applications) (Nauka, Moscow 1981)
8.31 H. Görtler: NASA Tech. Memo, p. 1375 (1954)
8.32 H. Bippes, H. Görtler: Acta Mech. **14**, 251 (1972)
8.33 S.A. Ragao, A.H. Nayfeh: Phys. Fluids **24**, 1405 (1981)
8.34 M. Dennefeld, S. Laustsen, J. Materne: Astron. Astrophys. **74**, 123 (1979)
8.35 I.I. Pasha: Astronomicheskiy Tsirkulyar **1387**, 4 (1985)
8.36 J.E. Beckman, S.G. Bransgrove, J.P. Phillips: Astron. Astrophys. **157**, 49 (1986)
8.37 V.L. Afanasyev, J. Boulesteix, F. Bonnarel, S.N. Dodonov, V.V. Vlasjuk: in Dynamics and Interaction of Galaxies, ed. by R. Wielen (Springer, Berlin, Heidelberg 1990), p. 354
8.38 G. Cantopulos: Astron. Astrophys. **64**, 323 (1978)
8.39 A.V. Zasov, O.K. Sil'chenko: Pis'ma v Astron. Zh. **13**, 455 (1987)
8.40 V.L. Afanasyev, A.N. Burenkov, A.V. Zasov, O.K. Sil'chenko: Astrofizika **28**, 243 (1988)
8.41 V.L. Afanasyev, A.N. Burenkov, A.V. Zasov, O.K. Sil'chenko: Astrofizika **28**, 1040 (1988)
8.42 V.L. Afanasyev, A.N. Burenkov, A.V. Zasov, O.K. Sil'chenko: Astrofizika **29**, 155 (1988)
8.43 S.G. Gestrin, V.M. Kontorovich: in Proc. Intern. Conf. on Plasma Phys. Kiev 1987, Contr. Papers, ed. by A.G. Sitenko (Naukova Dumka, Kiev 1987), Vol. 4, p. 245
8.44 S.G. Gestrin, V.M. Kontorovich: Pis'ma v Astron. Zh. **13**, 648 (1987)
8.45 Yu.N. Eliseev, K.N. Stepanov: Astrophys. and Space Sci. **114**, 233 (1985)

Chapter 9

9.1   S.V. Antipov, M.V. Nezlin, E.N. Snezhkin, A.S. Trubnikov: JETP Lett. **33**, 351 (1981)
9.2   S.V. Antipov, M.V. Nezlin, E.N. Snezhkin, A.S. Trubnikov: Sov. Phys. JETP **55**, 85 (1982)
9.3   S.V. Antipov, M.V. Nezlin, V.K. Rodionov, E.N. Snezhkin, A.S. Trubnikov: JETP Lett. **35**, 645 (1982)
9.4   M.V. Nezlin, E.N. Snezhkin, A.S. Trubnikov: JETP Lett. **36**, 234 (1982)
9.5   S.V. Antipov, M.V. Nezlin, V.K. Rodionov, E.N. Snezhkin, A.S. Trubnikov: Sov. Phys. JETP **57**, 786 (1983)
9.6   S.V. Antipov, M.V. Nezlin, E.N. Snezhkin, A.S. Trubnikov: in Nonlinear and Turbulent Processes in Physics, ed. by R.Z. Sagdeev (Gordon & Breach, New York 1984), Vol. 2, p. 665
9.7   S.V. Antipov, M.V. Nezlin, E.N. Snezhkin, A.S. Trubnikov: Sov. Phys. JETP **62**, 1097 (1985)
9.8   S.V. Antipov, M.V. Nezlin, E.N. Snezhkin, A.S. Trubnikov: Nature **323**, 238 (1986)
9.9   M.V. Nezlin, E.N. Snezhkin: in Nelineinye volny (Nonlinear Waves), ed. by A.V. Gaponov-Grekhov, M.I. Rabinovich (Nauka, Moscow 1987), p. 67
9.10 S.V. Antipov, M.V. Nezlin, V.K. Rodionov, A.Yu. Rylov, E.N. Snezhkin, A.S. Trubnikov, A.V. Khutoretsky: Sov. J. Plasma Phys. **14**, 648 (1988)

9.11   M.V. Nezlin, E.N. Snezhkin, A.S. Trubnikov: in Proc. Intern. Conf. on Plasma Phys. Kiev 1987, Inv. Papers, ed. by A.G. Sitenko (World Scientific, Singapore 1987), Vol. 2, p. 1184

9.12   M.V. Nezlin: Sov. Phys. Usp. **29**, 807 (1986)

9.13   J.G. Charney, G.R. Flierl: in Evolution of Physical Oceanography, ed. by B.A. Warren, C. Wunsch (MIT Press, Cambridge 1981)

9.14   G.R. Flierl: POLYMODE News **62**, 1 (1979)

9.15   E.N. Mikhailova, N.B. Shapiro: Izv. Atm. Ocean Phys. **16**, 587 (1980)

9.16   V.I. Petviashvili: Sov. Phys. JETP Lett. **32**, 619 (1980)

9.17   Yu.A. Danilov, V.I. Petviashvili: in Itogi nauki i tekhniki, ser. Fizika plasmy (Advances in Science and Technology, Ser. Plasma Physics) No.4, ed. by V.D. Shafranov (VINITI, Moscow 1983), p. 5

9.18   N.N. Romanova, V.Yu. Tseitlin: Izv. Atm. Ocean Phys. **20**, 85 (1984)

9.19   G.G. Sutyrin: Dokl. AN SSSR **280**, 1101 (1985)

9.20   G.G. Sutyrin: in Vnutritermoklinnye vikhri v okeane (Intrathermoclinic Vortices in the Ocean), ed. by K.N. Fedorov (Acad. Sci. USSR, P.P. Shirshov Institute of Oceanology, Moscow 1986), p. 93

9.21   G.G. Sutyrin: Izv. AN SSSR, Mekhanika Zhidkosti i Gaza **4**, 119 (1985)

9.22   G.G. Sutyrin: Dokl. AN SSSR **296**, 1076 (1987)

9.23   G.G. Sutyrin, I.G. Yushina: Dokl. AN SSSR **299**, 580 (1988)

9.24   G.G. Sutyrin, I.G. Yushina: Dokl. AN SSSR **288**, 585 (1986)

9.25   G.G. Sutyrin, I.G. Yushina: Izv. AN SSSR, Mekhanika Zhidkosti i Gaza **4**, 52 (1986)

9.26   B.A. Smith, R.A. Soderblom, T.V. Johnson, et al.: Science **204**, 951 (1979)

9.27   B.A. Smith, R.A. Soderblom, R. Batson, et al.: Science **215**, 504 (1982)

9.28   T. Matsuura, T. Yamagata: J. Phys. Oceanogr. **12**, 440 (1982)

9.29   G.P. Williams, T. Yamagata: J. Atm. Sci. **41**, 453 (1984)

9.30   G.P. Williams: Adv. Geophys. **28A**, 381 (1985)

9.31   G.P. Williams, R.J. Wilson: J. Atm. Sci. **45**, 207 (1988)

9.32   R.F. Beebe, T.A. Hockey: Icarus **65**, 86 (1986)

9.33   A.P. Ingersoll, R.L. Miller: Icarus **65**, 370 (1986)

9.34   M.M. McLow, A.P. Ingersoll: Icarus **65**, 353 (1986)

9.35   S.S. Limaye: Icarus **65**, 335 (1986)

9.36   A. Hatzes, D.D. Wenkert, A.P. Ingersoll, G.E. Danielson: J. Geophys. Res. **A86**, 8745 (1981)

9.37   B. Cushman-Roisin, B. Tang: in Mesoscale/Synoptic Coherent Structures in Geophysical Turbulence, ed. by J.C.J. Nihoul, B.M. Jamart (Elsevier, Amsterdam 1989), p. 51

9.38   Yu.L. Chernous'ko: Izv. Atm. Ocean Phys. **16**, 285 (1980)

9.39   F.V. Dolzhansky: Izv. Atm. Ocean. Phys. **17**, 413 (1981)

9.40   H. Greenspan: The Theory of Rotating Fluids (Cambridge University Press, Cambridge 1968)

9.41   U. Cederlöf: J. Fluid Mech. **187**, 395 (1988)

9.42   G.R. Flierl, M.E. Stern, J.A. Whitehead: Dyn. Atm. Ocean **7**, 233 (1983)

9.43   A.C. Scott, F.Y.E. Chu, D.W. McLaughlin: Proc. IEEE **61**, 1443 (1973)

9.44   V.E. Zakharov, S.V. Manakov, S.P. Novikov, L.P. Pitayevsky: Teoriya solitonov (Theory of Solitons) (Nauka, Moscow 1980)

9.45   K. Lonngren, A. Scott (eds.): Solitons in Action (Academic Press, New York 1978)

9.46   R.K. Bullough, P.J. Caudrey (eds.): Solitons (Springer, Berlin, New York, Heidelberg 1980)

9.47   R.K. Dodd, J.C. Eilbeck, J.D. Gibbon, H.C. Morris: Solitons and Nonlinear Wave Equations (Academic Press, London 1982)

9.48   P.G. Drazin: Solitons (Cambridge University Press, Cambridge 1983)

9.49   A.C. Newell: Solitons in Mathematics and Physics (Society for Industrial and Applied Mathematics, University of Arizona, Tucson 1985)

9.50   B.B. Kadomtsev: Kollektivnye yavleniya v plasme (Collective Phenomena in Plasma) (Nauka, Moscow 1988)

9.51   L.M. Degtyarev, V.G. Makhan'kov, L.I. Rudakov: Sov. Phys. JETP **40**, 264 (1974)

9.52   K. Lonngren: Plasma Phys. **25**, 943 (1983)

9.53 Y. Nakamura: IEEE Trans. Plasma Sci. **PS-10**, 180 (1982)
9.54 M.V. Nezlin: Physics of Intense Beams in Plasmas (IOP Publishing, Bristol, Philadelphia 1993)
9.55 M.V. Nezlin: in Itogi nauki i tekhniki, ser. Fizika plasmy (Advances in Science and Technology, Ser. Plasma Physics) No.5, ed. by V.D. Shafranov (VINITI, Moscow 1984), p. 5
9.56 M.V. Nezlin, G.G. Sutyrin: in Mesoscale/Synoptic Coherent Structures in Geophysical Turbulence, ed. by J.C.J. Nihoul, B.M. Jamart (Elsevier, Amsterdam 1989), p. 701
9.57 V.D. Larichev: Okeanologiya **23**, 551 (1983)
9.58 J. Nycander: in Proc. IV Intern. Workshop on Nonlinear and Turbulent Proc. in Physics, ed. by A.G. Sitenko, V.E. Zakharov, V.M. Chernousenko (Naukova Dumka, Kiev 1989) Vol. 1, p. 399
9.59 P.B. Rhines: Ann. Rev. Fluid Mech. **11**, 401 (1979)
9.60 G.P. Williams: Nature **257**, 778 (1975)
9.61 J. Nycander: J. Plasma Phys. **39**, 413 (1988)
9.62 P.S. Marcus: Nature **331**, 693 (1988)

Chapter 10

10.1 R. Hide: Nature **190**, 895 (1961)
10.2 H. Greenspan: The Theory of Rotating Fluids (Cambridge University Press, Cambridge 1968)
10.3 A.P. Ingersoll: Science **182**, 1346 (1973)
10.4 T. Maxworthy, L.G. Redekopp: Icarus **29**, 261 (1976)
10.5 T. Maxworthy, L.G. Redekopp: Science **210**, 1350 (1980)
10.6 L.G. Redekopp: J. Fluid Mech. **82**, 725 (1977)
10.7 T. Maxworthy, L.G. Redekopp, P.D. Weidman: Icarus **33**, 388 (1978)
10.8 D.N. Beaumont: Icarus **41**, 400 (1980)
10.9 B.A. Smith, R.A. Soderblom, T.V. Johnson et al.: Science **204**, 951 (1979)
10.10 B.A. Smith, R.A. Soderblom, R. Batson et al.: Science **215**, 504 (1982)
10.11 M.M. McLow, A.P. Ingersoll: Icarus **65**, 353 (1986)
10.12 R.Z. Sagdeev, V.D. Shapiro, V.I. Shevchenko: Sov. Astron. Lett. **7**, 279 (1981)
10.13 A.S. Volokitin, V.V. Krasnosel'skikh: Sov. Phys. Dokl. **26**, 863 (1981)
10.14 V.I. Petviashvili: Sov. Astron. Lett. **9**, 137 (1983)
10.15 T. Matsuura, T. Yamagata: J. Phys. Oceanogr. **12**, 440 (1982)
10.16 G.P. Williams, T. Yamagata: J. Atm. Sci. **41**, 453 (1984)
10.17 G.P. Williams: Adv. Geophys. **28A**, 381 (1985)
10.18 G.P. Williams, R.J. Wilson: J. Atm. Sci. **45**, 207 (1988)
10.19 M.V. Nezlin: JETP Lett. **34**, 77 (1981)
10.20 M.V. Nezlin: Sov. Astron. Lett. **10**, 221 (1984)
10.21 E.N. Mikhailova, N.B. Shapiro: Izv. Atm. Ocean Phys. **16**, 587 (1980)
10.22 G.R. Flierl: POLYMODE News **62**, 1 (1979)
10.23 V.I. Petviashvili: Sov. Phys. JETP Lett. **32**, 619 (1980)
10.24 M.V. Nezlin: Sov. Phys. Usp. **29**, 807 (1986)
10.25 P.S. Marcus: Nature **331**, 693 (1988)
10.26 J. Sommeria, S.D. Meyers, H.L. Swinney: Nature **331**, 689 (1988)
10.27 S.D. Meyers, J. Sommeria, H.L. Swinney: Physica **D37**, 515 (1989)
10.28 J. Sommeria, S.D. Meyers, H.L. Swinney: Nature **337**, 58 (1989)
10.29 J. Sommeria, S.D. Meyers, H.L. Swinney: in Nonlinear Topics in Ocean Physics, ed. by A. Osborn (North Holland, Amsterdam 1989)
10.30 L.A. Sromovsky, H.E. Revercomb, R.J. Krauss, V.E. Suomi: J. Geophys. Res. **A88**, 8650 (1983)
10.31 D.A. Godfrey, V. Moore: Icarus **68**, 313 (1986)
10.32 V.D. Larichev: Okeanologiya **23**, 551 (1983)

10.33 J.C. McWilliams: Rev. Geophys. **23**, 165 (1985)
10.34 A.G. Kostyanoi, I.M. Belkin: in Mesoscale/Synoptic Coherent Structures in Geophysical Turbulence, ed. by J.C.J. Nihoul, B.M. Jamart (Elsevier, Amsterdam 1989), p. 821
10.35 G.G. Sutyrin: in Vnutritermokl. vikhri v okeane (Intrathermoclinic Vortices in the Ocean), ed. by K.N. Fedorov (Acad. Sci. USSR, P.P. Shirshov Inst. of Oceanology, Moscow 1986), p. 93
10.36 S.V. Antipov, M.V. Nezlin, E.N. Snezhkin, A.S. Trubnikov: Sov. Phys. JETP **55**, 85 (1982)
10.37 M.V. Nezlin, G.G. Sutyrin: in Mesoscale/Synoptic Coherent Structures in Geophysical Turbulence, ed. by J.C.J. Nihoul, B.M. Jamart (Elsevier, Amsterdam 1989), p. 701
10.38 A.G. Kostyanoi, G.I. Shapiro: in Vnutritermoklinnye vikhri v okeane (Intrathermoclinic Vortices in the Ocean), ed. by K.N. Fedorov (Acad. Sci. USSR, P.P. Shirshov Inst. of Oceanology, Moscow 1986), p. 120
10.39 G.I. Shapiro: in Vnutritermoklinnye vikhri v okeane [as detailed in the previous reference] pp. 56, 71, 79 Vortices in the Ocean), ed. by K.N. Fedorov (Acad. Sci. USSR, P.P. Shirshov Inst. of Oceanology, Moscow 1986),
10.40 J.C. McWilliams, P.R. Gent, N.J. Norton: J. Phys. Oceanogr. **16**, 838 (1986)
10.41 S.E. McDowell, H.T. Rossby: Science **202**, 1085 (1978)
10.42 L. Armi, D. Hebert, N. Oakey: Nature **333**, 649 (1988)
10.43 J. Marshall: Nature **333**, 594 (1988)
10.44 G.I. Shapiro: in Mesoscale/Synoptic Coherent Structures in Geophysical Turbulence, ed. by J.C.J. Nihoul, B.M. Jamart (Elsevier, Amsterdam 1989), p. 783
10.45 D.C. Smith IV, R.O. Reid: J. Phys. Oceanogr. **12**, 244 (1982)
10.46 G.G. Sutyrin: Dokl. AN SSSR **296**, 1076 (1987)
10.47 G.G. Sutyrin, I.G. Yushina: Izv. AN SSSR, Mekhanika Zhidkosti i Gaza 4, 52 (1986)

Chapter 11

11.1 V.D. Larichev, G.M. Reznik: Dokl. AN SSSR **231**, 1077 (1976)
11.2 S.V. Antipov, M.V. Nezlin, E.N. Snezhkin, A.S. Trubnikov: Sov. Phys. JETP **55**, 85 (1982)
11.3 M.V. Nezlin: Sov. Phys. Usp. **29**, 807 (1986)
11.4 M.V. Nezlin, E.N. Snezhkin: in Nelineinye volny (Nonlinear Waves), ed. by A.V. Gaponov-Grekhov, M.I. Rabinovich (Nauka, Moscow 1987), p. 67
11.5 S.V. Antipov, M.V. Nezlin, V.K. Rodionov, A.Yu. Rylov, E.N. Snezhkin, A.S. Trubnikov, A.V. Khutoretsky: Sov. J. Plasma Phys. **14**, 648 (1988)
11.6 M.E. Stern: J. Marine. Res. **33**, 1 (1975)
11.7 G.R. Flierl, M.E. Stern, J.A. Whitehead: Dyn. Atm. Ocean **7**, 233 (1983)
11.8 J.A. Whitehead: in Mesoscale/Synoptic Coherent Structures in Geophysical Turbulence, ed. by J.C.J. Nihoul, B.M. Jamart (Elsevier, Amsterdam 1989), p. 627
11.9 R.W. Griffits, P.F. Linden: J. Fluid Mech. **105**, 283 (1981)
11.10 K.N. Fedorov, A.I. Ginzburg, A.G. Kostyanoi: in Mesoscale/Synoptic Coherent Structures in Geophysical Turbulence, ed. by J.C.J. Nihoul, B.M. Jamart (Elsevier, Amsterdam 1989), p. 15
11.11 Y. Couder, C. Basdevant: J. Fluid Mech. **173**, 225 (1986)
11.12 J.C. McWilliams, N.J. Zabusky: Geophys. Astrophys. Fluid Dynamics **19**, 207 (1982)
11.13 K.N. Fedorov, A.I. Ginzburg: in Mesoscale/Synoptic Coherent Structures in Geophysical Turbulence, ed. by J.C.J. Nihoul, B.M. Jamart (Elsevier, Amsterdam 1989), p. 1
11.14 A.G. Kostyanoi, I.M. Belkin: in Mesoscale/Synoptic Coherent Structures in Geophysical Turbulence, ed. by J.C.J. Nihoul, B.M. Jamart (Elsevier, Amsterdam 1989), p. 821
11.15 V.V. Dolotin, A.M. Fridman: in Nonlinear Waves, Physics and Astrophysics, ed. by A.V. Gaponov-Grekhov, M.I. Rabinovich (Springer, New York 1990)
11.16 R.C. Kloosterziel, G.J.F. van Heijst: in Mesoscale/Synoptic Coherent Structures in Geophysical Turbulence, ed. by J.C.J. Nihoul, B.M. Jamart (Elsevier, Amsterdam 1989), p. 609
11.17 R.C. Kloosterziel, G.J.F. van Heijst: J. Fluid Mech. **223**, 1 (1991)

Chapter 12

12.1   V.P. Pavlenko, V.V. Taranov: Fizika Plazmy **10**, 1303 (1984)
12.2   V.I. Petviashvili, A.P. Smirnov: Dokl. AN SSSR **277**, 88 (1984)
12.3   V.P. Pavlenko, V.V. Taranov: Fizika Plazmy **12**, 1329 (1986)
12.4   G.L. Lamb Jr.: Elements of Soliton Theory (Wiley, New York 1980)
12.5   K. Lonngren, A. Scott (eds.): Solitons in Action (Academic Press, New York 1978)
12.6   A. Hasegawa, K. Mima: Phys. Fluids **21**, 87 (1978)
12.7   A. Hasegawa, S.G. McLennan, Y. Kodama: Phys. Fluids **22**, 2122 (1979)
12.8   A. Hasegawa: Adv. Phys. **34**, 1 (1985)
12.9   M. Makino, T. Kamimura, T. Tainuti: J. Phys. Soc. Japan **50**, 990 (1981)
12.10 L.A. Mikhailovskaya: Fizika Plazmy **12**, 879 (1986)
12.11 N.S. Erokhin, L.A. Mikhailovskaya, V.M. Chernousenko: Preprint of Inst. Theor. Phys. (Kiev) No.85-81 (1985)
12.12 H.L. Pecseli, Y. Rasmussen, K. Thomsen: Phys. Rev. Lett. **52**, 2148 (1984)
12.13 H.L. Pecseli, Y. Rasmussen, K. Thomsen: Plasma Phys. **27**, 837 (1985)
12.14 S.I. Vainshtein: Magnitnaya Gidrodinamika **4**, 73 (1985)
12.15 J. Nycander: Phys. Scripta **39**, 758 (1989)
12.16 J. Nycander: Phys. Fluids **B1**, 1788 (1989)
12.17 J. Nycander: in Proc. IV Intern. Workshop on Nonlinear and Turbulent Proc. in Physics, ed. by A.G. Sitenko, V.E. Zakharov, V.M. Chernousenko (Naukova Dumka, Kiev 1989) Vol. 1, p. 399
12.18 J. Nycander: J. Plasma Phys. **39**, 413 (1988)
12.19 T. Crowley, E. Mazzucato: Nuclear Fusion **25**, 507 (1985)
12.20 TFR Group, A. Truc: Nuclear Fusion **26**, 1303 (1986)
12.21 P.C. Liever: Nuclear Fusion **25**, 543 (1985)
12.22 S.J. Zweben: Phys. Fluids **28**, 974 (1985)
12.23 S.V. Antipov, M.V. Nezlin, V.K. Rodionov, A.Yu. Rylov, E.N. Snezhkin, A.S. Trubnikov, A.V. Khutoretsky: Sov. J. Plasma Phys. **14**, 648 (1988)
12.24 J.D. Meiss, W. Horton: Phys. Fluids **26**, 990 (1983)
12.25 W. Horton, J. Liu, J.D. Meiss, J.E. Sedlak: Phys. Fluids **29**, 1004 (1986)
12.26 W. Horton: Phys. Fluids **B1**, 524 (1989)
12.27 P.L. Read, R. Hide: Nature **302**, 126 (1983)
12.28 P.L. Read, R. Hide: Nature **308**, 45 (1984)
12.29 M.A. Vlasov: Pis'ma v Zh. Eksp. Teor. Fiz. **2**, 274 (1965)
12.30 M.A. Vlasov: Pis'ma v Zh. Eksp. Teor. Fiz. **2**, 297 (1965)
12.31 E.I. Dobrokhotov, A.V. Zharinov, I.N. Moskalev, D.P. Petrov: Nuclear Fusion **9**, 143 (1969)
12.32 L.I. Elizarov, A.V. Zharinov: Nuclear Fusion, Suppl. pt. 2, 699 (1962)
12.33 W.H. Bostick: IEEE Trans. Plasma Sci. **PS-14**, 703 (1986)

Chapter 13

13.1   H. Haken: Advanced Synergetics (Springer, Berlin, Heidelberg, New York 1983)
13.2   I. Prigogine: From Being to Becoming (Freeman, San Francisco 1980)
13.3   H. Haken: Information and Self-Organization (Springer, Heidelberg 1988)
13.4   H. Haken (ed.): Synergetics (Springer, Berlin, Heidelberg, New York 1977)
13.5   B.B. Kadomtsev: Comm. on Plasma Phys. Contr. Fusion **11**, 153 (1987)
13.6   A.V. Gaponov-Grekhov, M.I. Rabinovich, J. Engelbrecht (eds.): Nonlinear Waves, Vols. 1,2 (Springer, Berlin, Heidelberg 1989)
13.7   V.I. Krinsky (ed.): Self-Organization: Autowaves and Structures Far from Equilibrium (Springer, Berlin, Heidelberg 1984)

13.8   M.I. Rabinovich, D.I. Trubetskov: Oscillations and Waves in Linear and Nonlinear Systems (Kluwer Academic Publishers, 1989)
13.9   I.S. Aranson, A.V. Gaponov-Grekhov, M.I. Rabinovich: Izv. AN SSSR, Fizika **51**, 1133 (1987)
13.10 A. Hasegawa: Adv. Phys. **34**, 1 (1985)
13.11 E.N. Parker: Astrophys. J. **162**, 665 (1970)
13.12 P.A. Gilman: Science **160**, 760 (1968)
13.13 A.S. Monin: Usp. Fiz. Nauk **132**, No. 1, 123 (1980)
13.14 E.J. Hopfinger, F.K. Browand, Y. Gagne: J. Fluid Mech. **125**, 505 (1982)
13.15 M.I. Rabinovich, M.M. Sushchik: Usp. Fiz. Nauk **160**, No. 1,3 (1990)
13.16 E.N. Pelinovsky: in [13.6], p. 128
13.17 D.R. Christie, K.J. Muirhead, A.L. Hales: J. Atm. Sci. **35**, 805 (1978)
13.18 L.A. Ostrovsky, Yu.A. Stepanyants: Rev. Geophys. **27**, 293 (1989)
13.19 R.A. Kerr: Science **245**, 929 (1989)
13.20 D. Lindley: Nature **341**, 95 (1989)
13.21 S.V. Antipov, M.V. Nezlin, E.N. Snezhkin, A.S. Trubnikov: Nature **343**, 517 (1990)
13.22 M.V. Nezlin, A.Yu. Rylov, A.S.Trubnikov, A.V. Khutoretsky: Geophys. Astrophys. Fluid Dynamics **52**, 211 (1990)
13.23 M.V. Nezlin: Izv. Atmosph. and Oceanic Phys. **27**, 22 (1991)
13.24 A.P. Ingersoll: Science **248**, 308 (1990)

Supplement S5

S5.1   B. Cushman-Roisin, B. Tang: Journ. Phys. Ocean. **20**, 97 (1990)
S5.2   A.B. Mikhailovskii: in Nonlinear Phenomena in Plasma Physics and Hydrodynamics, ed. by R.Z. Sagdeev (Mir, Moscow 1986), p. 8
S5.3   V.N. Oraevsky, H. Tasso, H. Wobig: in Plasma Physics and Controlled Nuclear Fusion Research (IAEA, Vienna 1969), Vol. 1, p. 671
S5.4   Yu.A. Danilov, V.I. Petviashvili: in Itogi nauki i tekhniki, ser. Fizika plazmy (Advances in Science and Technology, Ser. Plasma Physics) No. 4, ed. by V.D. Shafranov (VINITI, Moscow 1983), p. 5
S5.5   J.D. Meiss, W. Horton: Phys. Fluids **25**, 1838 (1982)
S5.6   T. Crowley, E. Mazzucato: Nuclear Fusion **25**, 507 (1985)
S5.7   TFR Group, A. Truc: Nuclear Fusion **26**, 1303 (1986)
S5.8   A.J. Wootton, B.A. Carreras, H. Matsumoto, K. McCuire, W.A. Peebles, Ch.P. Ritz, P.W. Terry, S.J. Zweeben: Phys. Fluids **B**, Plasma Phys., **2**, 2879 (1990)
S5.9   W. Horton: Preprint IFSR-416 (Institute for Fusion Studies, Texas Univ., Austin (1990)
S5.10 B.D. Scott, H. Biglari, P.W. Terry, P.H. Diamond: Phys. Fluids **3B**, 51 (1991)

Supplement S8

S8.1   L.S. Marochnik, A.A. Suchkov: Galaktika (Galaxy) (Nauka, Moscow 1984)
S8.2   J. Binney, S. Tremaine: Galactic Dynamics (Princeton Univ. Press 1987)
S8.3   A. Toomre: in Structure and Evolution of Normal Galaxies, ed. by S.M. Fall, D. Linden-Bell (Cambridge University Press, Cambridge 1981), p. 111
S8.4   A.A. Sumin, A.M. Fridman and U.A. Haud: Pis'ma v Astron. Zh. **17**, 698 (1991) [Sov. Astron. Lett.]

Supplement S10

S10.1 T.E. Dowling, A.P. Ingersoll: J. Atm. Sci. **46**, 3256 (1989)
S10.2 A.P. Ingersoll: Science **248**, 308 (1990)

S10.3 G.P. Williams, T. Yamagata: J. Atm. Sci. **41**, 453 (1984)
S10.4 F.H. Busse: PAGEOPH **121**, 375 (1983)
S10.5 J.I. Yano: Meteor. Soc. Japan **65**, 313 (1987)
S10.6 R. Hanel, B. Conrath, M. Flasar *et al.*: Science **206**, 952 (1979)
S10.7 B. Cushman-Roisin, B. Tang: Journ. Phys. Oceanogr. **20**, 97 (1990)
S10.8 P.S. Marcus, J. Sommeria, S.D. Meyers, H.L. Swinney: Nature **343**, 517 (1990)
S10.9 M.M. MacLow, A.P. Ingersoll: Icarus **65**, 353 (1986)
S10.10 B.A. Smith, R.A. Soderblom, T.V. Johnson, A.P. Ingersoll *et al.*: Science **204**, 927 (1979)

Additional Literature
Published After Completion of the Main Body of the Book

To Chapter 2

    Reviews on vortices in nature and laboratory
1   A.S. Monin, G.M. Zhikharev: Usp. Fiz. Nauk **160**, No. 5, 1 (1990)
2   F.V. Dolzhanskii, V.A. Krymov, D.Yu. Manin: Usp. Fiz. Nauk **160**, No. 7, 1 (1990)
    Original papers on vortices and solitons
3   J.C. McWilliams: Phys. Fluids **A2**, 547 (1990)
4   J.N. James: Nature **348**, 283 (1990)
5   M.K. Ramamurthy, B.P. Collins, R.M. Rauber, P.C. Kennedy: Nature **348**, 314 (1990)

To Chapter 10

1   J. Sommeria, C. Nore, T. Dumont, R. Robert: C. R. Acad. Sci. Paris **312**, No. 2, 999 (1991)
2   J. Sommeria, C. Staquet, R. Robert: J. Fluid Mech. (1991), in press

To Chapter 12

1   J. Nycander: Preprint UPTEC 90 003R (Uppsala Univ., Sweden 1990)
2   J. Nycander: Preprint UPTEC 90 045R (Uppsala Univ., Sweden 1990)

To Chapter 13

1   M.A. Goldstick, V.N. Stern, N.I. Yavorskii: Vyazkiye techeniya s paradoksalnymi svoistvami (Viscous Flows with Paradoxical Properties) (Nauka, Novosibirsk 1989)
2   A.V. Gaponov-Grekhov, M.I. Rabinovich: Physics Today, July 1990, p. 30
3   A.V. Gaponov-Grekhov, M.I. Rabinovich: Nonlinearities in Action (Springer, Berlin, Heidelberg, New York 1992)

Subject Index

B. N. Zakhariev, Moscow; **A. A. Suzko,** Minsk, USSR

Direct and Inverse Problems

Potentials in Quantum Scattering

1990. XIII, 223 pp. 42 figs. Softcover
ISBN 3-540-52484-3

This textbook can almost be viewed as a "how-to" manual for solving quantum inverse problems, that is, for deriving the potential from spectra or scattering data and also, as somewhat of a quantum "picture book" which should enhance the reader's quantum intuition. The formal exposition of inverse methods is paralleled by a discussion of the direct problem. Differential and finite-difference equations are presented side by side. The common features and (dis)advantages of a variety of solution methods are analyzed. To foster a better understanding, the physical meaning of the mathematical quantities are discussed explicitly. Wave confinement in continuum bound states, resonance and collective tunneling, energy shifts and the spectral and phase equivalence of various interactions are some of the physical problems covered.

D. Park, Williams College, Williamstown, MA

Classical Dynamics and Its Quantum Analogues

2nd enl. and updated ed. 1990. IX, 333 pp. 101 figs. Hardcover ISBN 3-540-51398-1

The primary purpose of this textbook is to introduce students to the principles of classical dynamics of particles, rigid bodies, and continuous systems while showing their relevance to subjects of contemporary interest. Two of these subjects are quantum mechanics and general relativity. The book shows in many examples the relations between quantum and classical mechanics and uses classical methods to derive most of the observational tests of general relativity. A third area of current interest is in nonlinear systems, and there are discussions of instability and of the geometrical methods used to study chaotic behaviour. In the belief that it is most important at this stage of a student's education to develop clear conceptual understanding, the mathematics is for the most part kept rather simple and traditional.
This book devotes some space to important transitions in dynamics: the development of analytical methods in the 18th century and the invention of quantum mechanics.

A. Hasegawa, AT & T Bell Laboratories, Murray Hill, NJ

Optical Solitons in Fibers

2nd enl. ed. 1990. XII, 79 pp. 25 figs.
Softcover ISBN 3-540-51747-2

Already after six months high demand made a new edition of this textbook necessary. The most recent developments associated with two topical and very important theoretical and practical subjects are combined: **Solitons** as analytical solutions of nonlinear partial differential equations and as lossless signals in dielectric **fibers.** The practical implications point towards technological advances allowing for an economic and undistorted propagation of signals revolutionizing telecommunications. Starting from an elementary level readily accessible to undergraduates, this pioneer in the field provides a clear and up-to-date exposition of the prominent aspects of the theoretical background and most recent experimental results in this new and rapidly evolving branch of science. This well-written book makes not just easy reading for the researcher but also for the interested physicist, mathematician, and engineer. It is well suited for undergraduate or graduate lecture courses.

Springer-Verlag
Berlin
Heidelberg
New York
London
Paris
Tokyo
Hong Kong
Barcelona
Budapest

A. V. Gaponov-Grekhov, M. I. Rabinovich,
Institute of Applied Physics, Gorky, USSR
(Eds.)

Nonlinearities in Action

1992. Approx. 200 pp. 180 figs. Hardcover
ISBN 3-540-51988-2

This textbook provides a concise and comprehensive overview on nonlinear problems addressing all natural scientists. It also contains a beautifully illustrated color insert easily accessible to interested laypersons. Thus it is suitable for leisure reading and also for an introductory (under-)graduate course in nonlinear physics. Both already well-known and some more recent new results are discussed, outlining the relation between classical problems of nonlinear physics and important current problems like the birth of chaos in simple deterministic systems and the birth of order out of disorder and turbulence.

Keywords: oscillations, nonlinear resonance, scattering, modulation, fractals, complexity, analysis of complex signals, Mandelstam scattering, Van der Pol oscillator, turbulent boundary flow on a plate, chaotic dynamics of lattice defects, particle-like solutions of three-dimensional fields, strange attractors.

Springer-Verlag
Berlin
Heidelberg
New York
London
Paris
Tokyo
Hong Kong
Barcelona
Budapest

Editors

Francesco Calogero

Dipartimento di Fisica
Università di Roma "La Sapienza"
Piazzale Aldo Moro 2
I-00185 Rome, Italy

Benno Fuchssteiner

Fachbereich Mathematik
Universität Paderborn
Warburgerstr. 100
D-33098 Paderborn, Germany

George Rowlands

Department of Physics
University of Warwick
Gibber Hill
Coventry, CV4 7AL
United Kingdom

Miki Wadati

Department of Physics
University of Tokyo
Hongo 7-3-1, Bunkyo-ku
Tokyo 113, Japan

Vladimir E. Zakharov

Landau Institute for Theoretical Physics
Russian Academy of Sciences
ul. Kosygina 2
117334 Moscow, Russia

Printed in Great Britain
by Amazon

53842537R00134